THE SOUL OF THE NIGHT

An Astronomical Pilgrimage

CHET RAYMO

Wood Engravings by
MICHAEL McCURDY

Prentice-Hall, Inc., Englewood Cliffs, New Jersey 07632

Library of Congress Cataloging in Publication Data

Raymo, Chet.
The soul of the night.

Includes index.
1. Astronomy. I. McCurdy, Michael. II. Title.
QB43.2.R39 1985   520   85-6480
ISBN 0-13-822883-3

First Edition

Printed in the United States of America

10 9 8 7 6 5 4 3 2 1

Book design by Alice R. Mauro
Jacket design and art direction by Hal Siegel
Jacket illustration by Michael McCurdy
Manufacturing buyer: Carol Bystrom

Prentice-Hall International (UK) Limited, *London*
Prentice-Hall of Australia Pty. Limited, *Sydney*
Prentice-Hall Canada Inc., *Toronto*
Prentice-Hall Hispanoamericana, S.A., *Mexico*
Prentice-Hall of India Private Limited, *New Delhi*
Prentice-Hall of Japan, Inc., *Tokyo*
Prentice-Hall of Southeast Asia Pte. Ltd., *Singapore*
Whitehall Books Limited, *Wellington, New Zealand*
Editora Prentice-Hall do Brasil Ltda., *Rio de Janeiro*

ISBN 0-13-822883-3

for Maureen

CONTENTS

PREFACE

Contemporary astronomy provides cautious answers to some of the roomiest questions humans ask about the universe. What *is* the universe? Where did it come from? How will it end? What is it made of? And what is this thing called life that dances on the surface of creation like an abiding flame?

The answers from the new astronomy present us with sweeps of space and time that seem wildly incommensurate with the human scale. Here are stars in uncountable numbers, each (perhaps) warming Earths unseen, other Earths teeming with other life. Here are galaxies where stars by the hundreds of billions are born in gassy nebulas and die in violence. Here are galaxies arrayed in knots and streamers across light-years, across *billions* of light-years, like motes of dust dancing in window light, worlds and worlds without end, reaching at last back to that singular moment when all that now exists came to be in a blinding flash of pure creation.

It is easy to be overawed by the visions of the new astronomy. Many among us would prefer to retreat into a comfortable cloud of unknowing. But if we are truly interested in knowing who we are, then we must be brave enough to accept what our senses and our reason tell us. We must enter into the universe of the galaxies and the light-years, *even at the risk of spiritual vertigo,* and know what after all must be known.

But to know is only half, as the naturalist John Burroughs said; to love is the other half. The pages that follow are an exercise in knowing and loving, a personal pilgrimage into the darkness and the silence of the night sky in quest of a human meaning. It is a quest rewarded with fleeting revelations, intimations of grace, and brief encounters with something greater than ourselves, a force, a beauty, and a grandeur that draw us into rapturous contemplation of the most distant celestial objects. And, occasionally, if we are lucky, the quest is rewarded with a special transcendent moment when the grandeur that abides in the night flares out (in the words of the poet Gerard Manley Hopkins) "like shining from shook foil."

The pilgrimage is one that each of us must make alone, into the realm of the stars and galaxies, to the limits of the universe, to that boundary of space and time where the mind and heart encounter the ultimate mystery, the known unknowable. It is a pilgrimage in quest of the soul of the night.

ACKNOWLEDGMENTS

Warm thanks to Robert Goulet and Anne Carrigg, who made important contributions to the book, and to Michael McCurdy, who graced the book with splendid illustrations. Mary Kennan, my editor at Prentice-Hall, was the book's attentive godparent. Eric Newman, also at Prentice-Hall, polished the manuscript with knowledge and skill. Alice Mauro and Hal Siegel gave the book a beautiful form. And a special thanks goes to Frank Ryan and Mike Horne, who were there when the pilgrimage began.

Page 6: The dedication is from Doris Lessing's *Shikasta,* Alfred A. Knopf, New York, 1979. *Page 8:* The line from Wallace Stevens is from the poem "Not Ideas About The Thing But the Thing Itself," from *The Collected Poems of Wallace Stevens,* Alfred A. Knopf, New York, 1973, and is used with permission. Copyright 1923, 1931, 1935, 1936, 1937, 1942, 1943, 1944, 1945, 1946, 1947, 1948, 1949, 1950, 1951, 1952, 1954 by Wallace Stevens. *Page 13:* The line from the poetry of Rainer Maria Rilke, as well as the lines on pages 145 and 147, are reprinted from *Duino Elegies,* by Rainer Maria Rilke, translated by J. B. Leishman and Stephen Spender, and are used by permission of W. W. Norton & Company, Inc., and The Horgarth Press Ltd. Copyright 1939 by W. W. Norton & Company, Inc. Copyright renewed 1967 by Stephen Spender and J. B. Leishman. *Page 14:* My characterization of the sun-hero is partly drawn from *Symbolism in Medieval Thought,* by Helen Flanders Dunbar, Russel & Russel, New York, 1961. *Page 17:* The line from the poetry of Theodore Roethke, as well as the lines on pages 20, 95, 192, 197, and 208 and the titles of Chapters 2, 9, and 19, are excerpted from "In a Dark Time" and "In Evening Air," copyright © 1960 by Beatrice Roethke as Administratrix for the Estate of Theodore Roethke from *The Collected Poems of Theodore Roethke.* Reprinted by permission of Doubleday & Co., Inc. *Page 25:* The quotation is from Vladimir Nabokov's *Lectures on Literature,* edited by Fredson Bowers, Harcourt Brace Jovanovich, New York, 1980. *Page 33:* The line from the poetry of Sylvia Plath ("This is the light . . . "), and those on pages 34 and 40, are excerpted from *The Collected Poems,* by Sylvia Plath, Harper & Row, New York, 1981. *Page 46:* The segments from the creation myth are excerpted from *Origins: Creation Texts from the Ancient Mediterranean,* co-edited and translated by Charles Doria and Harris Lenowitz, Anchor/Doubleday, Garden City, New York, 1976. *Page 95:* The work described here is *Five Kingdoms: An Illustrated Guide to the Phyla of Life on Earth*, by L. Margulis and K. V. Schwartz, W. H. Freeman & Co., San Francisco, 1982. *Page 101:* Here, as elsewhere in this book, I am indebted to Robert Burnham's *Robert Burnham's Celestial Handbook*, Dover, New York, 1978. *Page 109:* The description of the Great Kraken is from *In the Wake of Sea-Serpents*, by Bernard Heuvelmans, Hill & Wang, New York, 1968. *Page 131:* The work referred to is *Diana and Nikon* by Janet Malcolm, D. R. Godine, Boston, 1980. *Page 164:* The work referred to is Guy Ottewell's *Astronomical Calendar 1983*, c/o Dept. of Physics, Furman University, Greenville, S.C. 29613, 1983. *Page 177:* The lines from Pliny are excerpted from the Loeb Classical Library edition of Pliny's works, translated by H. R. Rackham, Harvard University Press, Cambridge, 1938. *Page 184:* The line from Robert Hass is excerpted from "Spring," *Field Guide*, by Robert Hass, Yale University Press, New Haven, 1973, and is used with permission. *Page 185:* The line from Wallace Stevens is excerpted from "Of Mere Being," *Opus Posthumous*, by Wallace Stevens, Alfred A. Knopf, New York, 1971. *Page 201:* All of the translations in this chapter from the poems of T'ao Ch'ien are from *Translations from the Chinese*, translated by Arthur Waley. Copyright 1919 and renewed 1947 by Arthur Waley. Reprinted by permission of Alfred A. Knopf, Inc., and George Allen & Unwin Ltd.

THE SOUL OF
THE NIGHT

THE SILENCE

Yesterday on Boston Common I saw a young man on a skateboard collide with a child. The skateboarder was racing down the promenade and smashed into the child with full force. I saw this happen from a considerable distance. It happened without a sound. It happened in dead silence. The cry of the terrified child as she darted to avoid the skateboard and the scream of the child's mother at the moment of impact were absorbed by the gray wool of the November day. The child's body simply lifted up into the air and, in slow motion, as if in a dream, floated above the promenade, bounced twice like a rubber ball, and lay still.

All of this happened in perfect silence. It was as if I were watching the tragedy through a telescope. It was as if the tragedy were happening on another planet. I have seen stars exploding in space, colossal, planet-shattering, distanced by light-years, framed in the cold glass of a telescope, utterly silent. It was like that.

During the time the child was in the air, the spinning Earth carried her half a mile to the east. The motion of the Earth about the sun carried her back again forty miles westward. The drift of the solar system among the stars of the Milky Way bore her silently twenty miles toward the star Vega. The turning pinwheel of the Milky Way Galaxy carried her 300 miles in a great circle about the galactic center. After that huge flight through space she hit the ground and bounced like a rubber ball. She lifted up into the air and flew across the Galaxy and bounced on the pavement.

It is a thin membrane that separates us from chaos. The child sent flying by the skateboarder bounced in slow motion and lay still. There was a long pause. Pigeons froze against the gray sky. Promenaders turned to stone. Traffic stopped on Beacon Street. The child's body lay inert on the asphalt like a piece of crumpled newspaper. The mother's cry was lost in the space between the stars.

How are we to understand the silence of the universe? They say that certain meteorites, upon entering the Earth's atmosphere, disintegrate with noticeable sound, but beyond the Earth's skin of air the sky is silent. There are no voices in the burning bush of the Galaxy. The Milky Way flows across the dark shoals of the summer sky without an audible ripple. Stars blow themselves to smithereens; we hear nothing. Millions of solar systems are sucked into black holes at the centers of the galaxies; they fall like feathers. The universe fattens and swells in a Big Bang, a fireball of Creation exploding from a pinprick of infinite energy, the ultimate firecracker; there is no soundtrack. The membrane is ruptured, a child flies through the air, and the universe is silent.

In Catholic churches between Good Friday and Easter Eve the bells are stilled. Following a twelfth-century European custom, the place of the bells is taken by *instruments des ténèbres* (instruments of darkness), wooden clackers and other noisemakers that remind the faithful of the terrifying sounds that were presumed to have accompanied the death of Christ. It was unthinkable that a god should die and the heavens remain silent. Lightning crashed about the darkened hill of Calvary. The veil of the temple was loudly rent. The Earth quaked and rocks split. Stars boomed in their courses. This din and thunder, according to medieval custom, are evoked by the wooden instruments.

Yesterday on Boston Common a child flew through the air, and there was no protest from the sky. I listened. I turned the volume of my indignation all the way up, and I heard nothing.

❦

There is a scene in Michelangelo Antonioni's film *Red Desert* in which a woman approaches a construction site where men are building a large linear-array radio telescope. "What is it for?" she asks. One of the workmen replies, "It is for listening to the stars." "Oh," she exclaims with innocent enthusiasm. "Can I listen?"

Let us listen. Let us connect the multimillion-dollar telescopes to our kitchen radios and convert the radiant energy of the stars into sound. What would we hear? The random crackle of the elements. The static of electrons fidgeting between energy levels in the atoms of stellar atmospheres. The buzz of hydrogen. The hiss and sputter of matter intent upon obeying the stochastic laws of quantum physics. Random, statistical, indifferent noise. It would be like the hum of a beehive or the clatter of shingle slapped by a wave.

In high school we did an experiment with an electric bell in a glass jar. The bell was suspended inside the jar, and the wires carrying electricity were led in through holes in the rubber stopper that closed the jar's mouth. The bell was set clanging. Then the air was pumped from the jar. Slowly, the sound of the bell was snuffed out. The clapper beat a silent tattoo. We watched the clapper thrashing silently in the vacuum, like a moth flailing its soft wings against the outside of a window pane.

Even by the standard of the vacuum in the bell jar, the space between the stars is empty. The emptiness between the stars is unimaginably vast. If the sun were a golf ball in Boston, the Earth would be a pinpoint twelve feet away, and the nearest star, Alpha Centauri, would be another golf ball (two golf balls, really, two golf balls and a pea; it is a triple star) in Cincinnati. The distances between the stars are huge compared with the sizes of the stars: a golf ball in Boston, two golf balls and a pea in Cincinnati, a marble in Miami, a basketball in San Francisco. The trackless trillions of miles between the stars are a vacuum more perfect than any vacuum that has yet been created on Earth. In our part of the Milky Way Galaxy, interstellar space contains about one atom of matter per cubic centimeter, one atom in every volume of space equal to the size of a sugar cube. The silent vacuum of the bell jar was a million times inferior to the vacuum of space. In the almost perfect vacuum of interstellar space, stars detonate, meteors blast craters on moons, and planets split at their seams with no more sound than the pulsing clapper of the bell in the evacuated jar.

Once I saw the Crab Nebula through a powerful telescope. The nebula is the expanding debris of an exploded star, a wreath of shredded star-stuff eight light-years wide and 5000 light-years away. What I saw in the telescope was hardly more than a blur of light, more like a smudge of dust on the mirror of the scope than the shards of a dying star. But seeing through a telescope is 50 percent vision and 50 percent imagination. In the blur of light I could easily imagine the outrushing shock wave, the expanding envelope of high-energy radiation, the torn filaments of gas, the crushed and pulsing remnant of the skeletal star. I stood for a quarter of an hour with my eye glued to the eyepiece of the scope. I felt a powerful sensation of energy unleashed, of an old building collapsing onto its foundations in a roar of dust at the precise direction of a demolition expert. As I watched the Crab Nebula, I felt as if I should be wearing earplugs, like an artilleryman or the fellow who operates a jackhammer. But there was no sound.

The Chinese saw the Crab when it blew up. In A.D. 1054 a new star appeared in Taurus. For weeks it burned more brightly than Venus, bright enough to be seen in broad daylight. Then the star gradually faded from sight. The Chinese recorded the "guest star" in their annals. Nine hundred years later the explosion continues. We point our telescopes to the spot in Taurus where the "guest star" appeared in 1054, and we see the bubble of furious gas still rushing outward.

Doris Lessing began her fictional chronicle of space with this dedication: *For my father, who used to sit, hour after hour, night after night, outside our home in Africa, watching the stars. "Well," he would say, "if we blow ourselves up, there's plenty more where we came from."* Yes, there's plenty more, all right, even if one or two blow themselves up now and then. A billion billion stars scattered in the vacuum of space. A star blew up for the Chinese in 1054. A star blew up for Tycho Brahe in 1572, and another for Kepler in 1604. They go in awesome silence.

❦

The physical silence of the universe is matched by its moral silence. A child flies through the air toward injury, and the galaxies continue to whirl on well-oiled axes. But why should I expect anything else? There are no Elysian Fields up there beyond the seventh sphere where gods pause in their revels to glance down aghast at our petty tragedies. What's up there is just one galaxy after another, magnificent in their silent turning, sublime in their huge indifference. The number of galaxies may be infinite. Our indignation is finite. Divide any finite number by infinity and you get zero.

Only a few hundred yards from the busy main street of my New England village, the Queset Brook meanders through a marsh as apparently remote as any I might wish for. To drift down that stream in November is to enter a primeval silence. The stream is dark and sluggish. It pushes past the willow roots and the thick green leaves of the arrowhead like syrup. The wind hangs dead in the air. The birds have fled south. Trail bikes are stacked away for the winter, and snowmobiles are still buried at the backs of garages. For a few weeks in November the marsh near Queset Brook is as silent as the space between the stars.

How fragile is our hold on silence. The creak of a wagon on a distant highway was sometimes noise enough to interrupt Thoreau's reverie. Thoreau was perceptive enough to know that the whistle of the Fitchburg Railroad (whose track lay close by Walden Pond) heralded something more than the arrival of the train, but he could hardly have imagined the efficiency with which technology has intruded upon our world of natural silence. Thoreau rejoiced in owls; their hoot, he said, was a sound well suited to swamps and twilight woods. The interval between the hoots was a deepened silence suggesting, said Thoreau, "a vast and undeveloped nature which men have not recognized." Thoreau rejoiced in that silent interval, as I rejoice in the silence of the November marsh.

As a student, I came across a book by Max Picard called *The World of Silence*. The book offered an insight that seems more valuable to me now than it did then. Silence, said Picard, is the source from which language springs, and to silence language must constantly return to be recreated. Only in relation to silence does sound have significance. It is for this silence, so treasured by Picard, that I turn to the marsh near Queset Brook in November. It is for this silence that I turn to the stars, to the ponderous inaudible turning of galaxies, to the clanging of God's great bell in the vacuum. The silence of the stars is the silence of creation and re-creation. It is the silence of that which cannot be named. It is a silence to be explored alone. Along the shore of Walden Pond the owl hooted a question whose answer lay hidden in the interval. The interval was narrow but infinitely deep, and in that deep hid the soul of the night.

I drift in my canoe down the Queset Brook and I listen, ears alert, like an animal that sniffs a meal or a threat on the wind. I am not sure what it is that I want to hear out of all this silence, out of this palpable absence of sound. A scrawny cry, perhaps, to use a phrase of the poet Wallace Stevens: "A scrawny cry from outside . . . a chorister whose c preceded the choir . . . still far away." Is that too much to hope for? I don't ask for the full ringing of the bell. I don't ask for a clap of thunder that would rend the veil in the temple. A scrawny cry will do, from far off there among the willows and the cattails, from far off there among the galaxies.

❦

The child sent flying by the young man on the skateboard bounced on the pavement and lay still. The pigeons froze against the gray sky. Promenaders turned to stone. How long was it that the child's body lay there like a piece of crumpled newspaper? How long did my heart thrash silently in my chest like the clapper of a bell in a vacuum? Perhaps it was a minute, perhaps only a fraction of a second. Then the world's old rhythms began again. A crowd gathered. Someone lifted the injured child into his arms and rushed with the mother toward help. Gawkers milled about distractedly and dispersed. The clamor of the city engulfed the Common. Traffic moved again on Beacon Street.

2

IN A DARK
TIME

O nce, when I was very young, my father woke me in the dead of night to see a comet. He had heard on the radio that a comet would be visible in the eastern sky in the hours before the dawn. Slippered and jacketed, dragging sleep behind me like a comet's tail, I followed my father into the yard. Together we stood among the black pines and searched our little patch of starry sky.

It is clear now, in the light of that memory, that my father did not know exactly what it was we were looking for or where in the sky we might find it. He imagined, I suppose, that the comet would announce itself, trumpeting like an angel, trailing a train of light. He imagined swoosh and glitter. He expected a sky on fire, and he wanted me to see it.

We did not see the comet. It was probably one of those dozen-a-year comets of the astronomers, visible only with binoculars or telescope or photographic plate. Or perhaps it was a faint naked-eye comet hidden from us by the pines. We did not see it. We stood in the frosty air and searched the sky until dawn lighted the east. I carry from that night my first memory of the stars, nameless, uncountable, flung like a cold net across the pines, beautiful and frightening.

❦

For Beauty's nothing but the beginning of Terror, laments the poet Rainer Maria Rilke. And tonight again I wake in darkness, tangled in the child's dream of crusty black pines and nameless stars. It is the solitary hour, the desperate hour, the hour of the wolf. Ghosts are harboring in the shadows, ghosts of the body, ghosts of the spirit. On an impulse, I rise and make my way through unlighted rooms to the dooryard. There, I catch the winter constellation Orion sneaking late across the autumn sky.

Giant Orion: boaster, beast-slayer, storm-bringer. Muscle-bulging constellation of the lights. No other part of the night sky visible to northern observers contains more brilliant stars. Here is diamond Rigel, the giant's forward foot, and ruby Betelgeuse, in the shoulder of the raised arm. Here are glittering Saiph and Bellatrix, the other foot and other shoulder, and Alnitak, Alnilam, and Mintaka, the white pearls of the hunter's belt. Bellatrix is the closest of these stars, but at 470 light-years away it is a hundred times more distant than our nearest stellar neighbors. Saiph is 2000 light-years away. These are the giants of the Galaxy, stars ten thousand times brighter than our sun, the largest and most intrinsically luminous stars of the night. These are the night answering the black pines with tongues of flame.

My father taught me to recognize and name the stars of Orion. He stood with me under the pines and pointed to the figure of a giant tangled and caught in the boughs of the trees. He carried a book, a book with star maps and star facts—and star stories. The story of Orion was the best of all.

> *After many adventures, Orion voyaged to the island of Chios, where he fell in love with Merope, the daughter of King Oenopion. The king agreed that Orion should wed his daughter but insisted that the giant should prove his worth by accomplishing a series of difficult tasks. As Orion completed each assignment, the king imposed another, each more demanding than the one that went before. At last Orion began to suspect that Oenopion had no intention of giving up his daughter and that the tasks would have no end. He decided to be done with it; he would carry Merope away by force. But the giant's plan was discovered by the king. Oenopion had Orion seized, blinded, and cast upon the strand by the water's edge to wander the Earth in darkness.*

Orion's story is part of a saga of light and darkness that flows like a sourceless stream through the folk memory of many cultures. The hero of the saga is invariably a warrior, brave and beautiful, mantled in clouds. His garments are dazzling. He loves a radiant maiden, whom he forsakes or slays. He is a traveler and a destroyer of beasts and monsters harmful to the fertility of the land. He is the "hero with a thousand faces" who journeys into darkness and effects a triumphant return.

The archetype for the Orion myth was the sun. The myth gave flesh to the primal human experience of the sun's seasonal and diurnal cycles—winter/summer, night/day, death and transfiguration. In the dreams of the early storytellers, blind Orion was the sun in winter or in night. The giant risked greatly, and now he walks in darkness. He stumbles westward, blinded and alone, abandoned by the gods. *Deus absconditus.* The gods have put out the lights. The gods have put out the lights and gone home. Orion is the Psalmist in the valley of shadow. He is John of the Cross in the soul's dark night. He is Dante at the dark wood, when the right road was wholly lost and gone.

The night is the beginning of terror, as every child knows. Who is not afraid of the dark? The gods are creatures of daylight. The gods work nine to five. At night we are on our own.

🦂

In the twenty minutes I stand watching in the dooryard, the stars of Orion move three finger-breadths toward their western setting. Clouds thicken and scud east. In the gaps between the racing clouds I glimpse the giant—there! the unmistakable stars of the belt!—halting, like the figure in a zoetrope. On the star maps in my father's book, the figure of Orion was displayed in his full powers, armed with a club and a lionskin shield, a sword dangling at his belt. But that confident image is not what I see this night of the fitful dream. Tonight it is the *blinded* giant I see, the pitiable, beautiful giant, the doomed son of Poseidon, stumbling, palms outstretched, westward across a wine-dark sea.

In other times, the brightest star in a constellation was called its *lucida*. Betelgeuse, in the shoulder of the giant, is the *lucida* of Orion. Blood red, like a blinded eye. It is a red supergiant star, a star bloated with fatigue. The diameter of Betelgeuse is 400 million miles. If Betelgeuse were where our sun is, the Earth and its orbit would be inside it. Mars, too, would be inside it. Betelgeuse is an old star that has puffed up to swallow its planets.

Giant Betelgeuse is one of the few stars that astronomers have succeeded in photographing as more than mere points of light. The star shows up in these unusual images as a luminous disk enclosed in an aura of gas, gas blown away from the surface of the star by its stellar wind. I've seen computer-enhanced false-color photographs of Betelgeuse. The colors are coded to represent temperatures on the star's surface. They indicate convective structures in the body of the star, great billowings of energy in the star's outer layers, currents stirred up by the furnace at the star's core. Orange "oceans" of color on the photographs mark the places where heat rises from the star's interior and breaks onto the surface. Blue "continents" are cooler regions where energy sinks back into the star. Who could have guessed it? Who could have guessed that those cold points of light in the night sky were other suns, suns blazing with the thermonuclear fires of Creation? Who could have guessed that the red star in Orion's shoulder outshines the noon?

These messages from the astronomers, these computer-enhanced postcards, have only recently arrived on our doorstep to tell us that Betelgeuse is a distended giant star, a sphere of hydrogen and helium 400 million miles in diameter, swelling with a violence that consumes planets and lights up dusky corners of the Galaxy. Its surface roils like a sea in storm, and a froth of fire is flung ten million miles into space. In Earth's sky this monstrous object is reduced to a point of light at Orion's shoulder. The stars are good at hiding their true natures. They have a trick up their sleeve, and that trick is distance. At the distance of Betelgeuse—500 light-years—an average-sized star like our sun would be invisible to the eye. Astronomers on the Palomar Mountain of a Betelgeusian planet might, if they were lucky, pick out our sun as a tiny speck on a photographic plate, a nondescript star among billions of Milky Way stars, a needle of light in the galactic haystack. Betelgeuse is conspicuous in our own night sky only because of its prodigious size.

❧

Behind thickening clouds, Orion stumbles west. The son of Poseidon, he walks upon the water. In his blindness, does he put one foot before the other with caution, afraid that he will step off the edge of the world? Do the fish become still in their salty pools as the giant timidly treads upon their starry roof? *In a dark time the eye begins to see:* The line is from a poem of Theodore Roethke. Isaiah has it too: *The people who walked the darkness have seen the light.* Medieval mystics embraced the darkness; only by passing through the dark night did they hope to attain the perfect light of the Beatific Vision.

I will settle for something less than the beatific. Like a fish in the dark sea, I will be content with the shadow of the giant who walks upon the water. The light of science is more cautiously received than the light of the mystics, and this is a book about science. This is a book about a vision of the universe that has been distilled by reason and imagination from faint starlight. This is a book about a new cosmology condensed from the night like drops of dew.

The light of science may be more cautiously received than the light of the mystics, but the quest is no less heroic. There are visions in the new astronomy grand enough to blow the gods from their celestial thrones. In the constellation Orion alone, there are sights and mysteries that would level Olympus. Stars are born there by the dozens in dusky nebulas, turning on with a thermonuclear brilliance that blinds planets. Stars like Rigel in Orion's foot, that burn with a keen blue light and exhaust in a mere million years more fuel than is contained in a hundred suns. Stars like the red giant Betelgeuse, bloated stars struggling to stave off a final gravitational collapse. Stars that heave and sigh a labored breath. Stars that explode. Stars that contract in their dying to the size of a planet, to the size of a city, even, and then—leaving behind a city-sized sphere of permanent blackness—shrink to the size of the head of a pin, and shrink still further until a mass equivalent to that of a dozen suns is compressed beyond physical dimension.

In the Horsehead Nebula, a black wisp of smoke near Orion's belt, there is room enough for 20,000 solar systems. Who would enter that dark wood? The Horsehead stands out against a backdrop of luminous gas, a sea of space resplendent with the rosy light of radiant hydrogen. Who would recognize his own face in that infinity? Neither the Horsehead nor its luminous backdrop is visible to the unaided eye. It is said that under exceptional conditions of darkness and clarity the Horsehead Nebula can be seen in a moderate-sized telescope equipped with a wide-field eyepiece. I have looked for it, but I have never seen it. I know this part of the sky only through the magnificent color photographs prepared by major observatories. It is a region of unparalleled riches: dazzling stars, dark nebulas, and shimmering banks of gas excited to luminescence by the radiation of huge Alnitak, the eastern star of Orion's belt. If there is something in the night akin to the visions of the mystics, this is it.

Wresting the secrets of the new cosmology from starlight is no longer the work of the hero with a thousand faces; it is the work of a thousand heroes with a single face—the community of science. But the quest is the same one that sent Orion stumbling across the wine-dark sea. *The night is our window on the Infinite.* We stumble through that darkness toward Light.

> *Orion heard the ringing of a blacksmith's hammer. Across the sea he followed the musical sound until he came at last to the island of Lemnos and the forge of the smith Hephaestus, the same Hephaestus, master craftsman, who fashioned moons of silver and suns of gold for the gods. Taking pity on blinded Orion, Hephaestus lent one of his own servants, Cedalion, to act as a guide. Together Cedalion and Orion set out for the east, the blind giant carrying the smith's servant on his broad shoulders, toward the land of Apollo and the rising sun. As Orion stood facing east, the sun rose. He felt the rays of the sun warm his eyes. And then he began to see, first as through a fog, and then as clearly as ever before.*

❦

The Greeks believed that the eye had a double role in vision. They believed that a pale light went out from the eye to the world and returned again to the eye as a traveler returns bearing gifts. For the Greeks, the eye was both illuminator and receiver. Modern science has rejected the Greek theory of vision. We are told that the eye is a passive agent, a mere collector of whatever light comes its way. Seeing, in the new dispensation, means turning toward the light and nothing more.

But there must be more to it than that, more than merely turning toward the light. Or why is it that we see most intensely during the dark times? A few days ago in the fading light of late afternoon, I came upon a clump of common joe-pye weed at the water-meadow. The plant burned with a purple flame. On a silver web between the stems of the purple plant an argiope spider turned its yellow belly-stripe toward the lowering sun, and color cut across the meadow like a bolt of lightning.

In a dark time the eye begins to see. And this is the paradox: that black is white, that darkness is the mother of beauty, that the extinction of light is a revelation. Were the Greeks right, then, after all? Perhaps it is only in the dark times that the pale light of intelligence, going out from the eye, can make its way in the world without being washed away by the fierce light of the sun. Perhaps it is only in the dark times that the eye and the mind, turning to each other, can cooperate in the delicate and impassioned art of seeing. Few people willingly choose to walk the dark path, to enter the dark wood, to feel the knot of fear in the stomach, or to live in the black cave of the sleepless night. But then, unexpectedly, the Greek truth emerges. The light of the mind returns bearing extraordinary gifts. "A man goes far to find out what he is," says the poet Theodore Roethke. An hour passed in the water-meadow. The argiope spider on its slender thread pivoted its yellow belly-stripe, like the hour hand of a clock, to follow the setting sun. "I meet my shadow in the deepening shade," continues Roethke. "The day's on fire. A steady stream of correspondences. A night flowing with birds, a ragged moon, and in broad day the midnight come again."

3

FAINT LIGHTS

O n March 10, 1982, I waited for the world to end. On that day, according to certain elements of the popular press, all of the planets in the solar system were due to line up on one side of the sun, like pieces of meat on a skewer. The predicted consequences of this rare alignment would be dramatic. The combined gravitational tug of the planets, pulling, as it were, on a single rope, would raise storms on the sun and havoc on Earth. Volcanoes would erupt along the Earth's stressed seams. Earthquakes would shake the continents. Geologic faults would open like zippers. California would fall into the sea.

Well, of course, it didn't happen. California is still zipped up along the San Andreas Fault. The sun still shines with a steady light, and the Earth still spins sedately on its old axis.

The planets did not line up at all, at least not like the diagrams in the papers, not like beads strung on an invisible string. At their most compact alignment, the planets occupied a pie-shaped wedge of 95 degrees as seen from the sun, which means they were spread out over more than one quarter of the sky. They had been very close to that same degree of alignment for weeks before the predicted calamity, with no noticeable effect. Furthermore, the planets had been lined up more closely than this many times in the Earth's history without consequence. And finally, the total gravitational excess of this particular planetary alignment could be precisely calculated by astronomers; jump from a second-story window onto the sidewalk and you will give the earth a more severe shaking.

And yet . . . to have all of the planets in the same quarter of the sky is a rare enough thing. A concentration of planets as tight as the alignment of 1982 occurs only once every 179 years. If you got up before the sun during the week of March 10, you had a chance to see all five naked-eye planets in the sky at the same time.

I was in my college's observatory at five o'clock on the morning of the predicted disasters. The sky was still inky black and ablaze with stars. I took the telescope and went from west to east on a grand tour of the solar system: Mars, Saturn, Jupiter, Venus, and Mercury. Even the moon graced my morning by placing her elegant crescent in the parade of planets. Saturn, wearing its rings like a jaunty cap, looked like a Saturn should, the archetypal cartoon planet, unlikely but undeniably real. Jupiter presented all four of its larger moons for my appreciation. Venus showed a thin crescent, mimicking the moon. Little Mercury, as usual, played hard-to-get. By the time Mercury stood above the treetops in the east the sky was turning pink. I climbed up to sit in the aperture of the observatory dome and watch the sun rise. The planets burned with an unusual intensity against the lightening sky. Blue Spica and red Antares added their twinkling brilliance to the stately row of planets. It was worth waiting 179 years for just such a morning.

❧

The world did not end on March 10, 1982. But the sky presented a feast of faint lights that will stay long in my memory. There were no earthquakes to knock me from my bed that morning, or to shake me from my roost in the aperture of the observatory dome. The sun did not erupt in streamers of flame. Instead, the sky whispered. The sky spoke sweet nothings. In my part of the world, near the lights and haze of a great city, it is hard enough to hear the sky whisper. I'll take every chance I get.

On the darkest, clearest nights thousands of stars are visible to the naked eye. In addition to stars, there are other naked-eye wonders available to the careful observer who is far from city lights: star clusters, at least one galaxy, nebulas, the Milky Way, the zodiacal light. But the typical urban or suburban observer might see only a few hundred of the brightest stars, and none of the more elusive objects. We have abused the darkness. We have lost the faint lights.

Once I saw the zodiacal light, the "false dawn" of Omar Khayyám. On clear moonless nights in August or September, an hour or two before dawn, the zodiacal light can be seen stretching up from the horizon along the band of the zodiac. The light is fainter than the Milky Way, and I had often looked for it unsuccessfully. The zodiacal light is caused by sunlight reflecting from interplanetary dust that populates the inner part of the solar system. The dust is mostly confined to a flat disk near the sun, which is why we see the light only just before sunrise or just after sunset, and only at those times of the year when the plane of the solar system is steeply tipped to the horizon. When at last I saw the zodiacal light, it was from a dark hillside near the sea. A wind from the Atlantic had cleared the air, and the stars walked on the horizon beneath a tall canopy of light. I never saw it again.

"Let us worship the spine and its tingle," Vladimir Nabokov advised students in his lectures on literature. "That little shiver behind is quite certainly the highest form of emotion that humanity has attained when evolving pure art and pure science." Searching for faint lights in the night sky is both an art and a science, and I count as worth a king's ransom the tingle in the spine that invariably accompanies a rare find. "We are vertebrates," said Nabokov, "tipped at the head with a divine flame." The brain is a continuation of the spine, an accretion of tissue at the top that burns with a pure blue flame, but the wick runs the whole length of the candle. The morning I saw the zodiacal light, I felt the heat of the flame all along the wick.

For another special feast of faint lights, try counting the Pleiades, the little cluster of stars in Taurus. Since antiquity these stars have been called the Seven Sisters, the Seven Virgins, or the Starry Seven. A poem of the third century B.C. names them: Alcyone, Merope, Celaeno, Taygeta, Sterope, Electra, and Maia, "small alike and faint." Most modern observers see only six Pleiades with the naked eye, or, if the night is exceptionally clear, then nine or ten. Maestlin, the tutor of Kepler, mapped eleven stars in the cluster prior to the invention of the telescope. The claim of twelve naked-eye Pleiades is not uncommon in the literature of the stars. The highest claim I have come across is sixteen. There are possibly as many as 500 stars that belong to the cluster. Twenty of these may have magnitudes that place them within the range of the naked eye, but the crowded massing of the stars makes even the claim of sixteen seem doubtful. The best I have done, on the darkest and clearest of nights, was nine. That was when I was younger and had better vision.

Or try looking for the young moon. Strictly, the moon is new at the moment it passes between the sun and the Earth. That moment can fall at any time of the day or night. What we call a "new moon" is in fact a young moon, perceived as a thin crescent at some time after the moon is exactly new. The youngest moons will be seen at sunset, with the crescent low in the west, near the place where the sun has just set, and distilled from the sun's glare by the horizon. Many factors—latitude, the time of sunset, the angle of the moon's orbit to the horizon—determine the likelihood of seeing a very young moon. Weather, the clarity of the horizon, and the keenness of the eye are also part of the game. I have seen a moon no older than thirty hours. It was as thin as the paring of a nail, as thin as an eyelash, crooked like Cupid's bow, its arrow aimed at the sun. But thirty hours is not a record, nor close to it. Twenty-four-hour-old moons are commonly reported. Lizzie King and Nellie Collinson, two housemaids at Scarborough, England, seem to have the record with a 14½-hour-old moon viewed on May 2, 1916. There is a legend that Johannes Kepler, the Renaissance astronomer, saw the old moon one morning and the young moon on the evening of the same day, but the story is hardly credible. I will go on looking. I will check my almanac and the *Astronomical Calendar* prepared annually by Guy

Ottewell. I will wait for that special evening when all of the factors click and the horizon is clear and the moon turns a shoulder toward me more slimly lit than any I have seen before, the one perfect lunar shrug.

There are other elusive feasts for the naked eye, faint lights in the night sky that belie their grandeur. The Great Galaxy in Andromeda is a difficult but accessible naked-eye object. I have discerned it on many a dark country night. Andromeda's misty blur was on star charts long before the invention of the telescope. No one guessed that the blur of light was an island universe of a trillion stars, another Milky Way Galaxy, the only galaxy other than our own that can be reliably seen without optical aid. The Andromeda Galaxy is 2 million light-years away. No older light will enter your eye without benefit of a telescope.

I have glimpsed M13, the globular cluster of a million stars in the constellation Hercules, or perhaps I only imagined that I saw it. Someday I will travel south and view the brighter globular clusters in the constellations Centaurus and Toucan. It is said that on the darkest nights it is possible to see the North American Nebula in Cygnus the Swan. The nebula is a patch of pale pink light crowded with stars, tucked beneath the swan's wing. I have often looked for it, but I have never seen it. I shall go on trying.

I have seen the Praesepe, the "beehive" of stars in Cancer that appears as a smudge of milky light to the naked eye. Hipparchus called it the "little cloud." When Galileo turned his telescope upon the "cloud," he was astonished and delighted to see it resolved into tiny diamondlike stars. He counted thirty-six. There are actually hundreds of stars in the cluster, twinkling beyond the limits of vision.

❧

I first saw the Praesepe from a hillside in the Catskill Mountains. Somehow I had made my way that day to the grave of John Burroughs, the turn-of-the-century American naturalist. There were two famous naturalists named John at the turn of the century. One was John Muir, "John-o'-the-Mountains," a man at home on alpine peaks and Alaskan glaciers, a man who took continents in his rangy stride. The other was John Burroughs, "John-o'-the-Birds," who was content to sit on his front porch and let the turning Earth bring all things to his door. Burroughs was a connoisseur of nature's subtle lights and whispered revelations. "The good observer of nature exists in fragments," he said, "a trait here and a trait there." Or again: "One secret of success in observing nature is [a] capacity to take a hint." Burroughs was a man who took the hints and trusted the tingle in his spine.

The site of Burroughs' grave lies at the top of a rocky meadow with splendid views across a broad valley. The grave is surrounded by a low wall of native shale. There is a spring nearby. Above the spring is the flat boulder where Burroughs came to sit and watch and take from the day its hints and traits. On the face of the boulder is fixed a plaque bearing the naturalist's epitaph: "John Burroughs / 1837–1921 / 'I stand amid the eternal ways and what is mine shall know my face.' "

Up on that hillside that spring morning there was no sound but that of the wind and the clacking of birches. I scared up a woodcock at the base of the boulder; it shot away in a whirring beeline, its long beak skimming the ground. This was clearly the sort of place that could produce a philosopher from a million partial traits and glimmers of truth. Within a bowl of tree-covered mountains, quiet meadows sloped down to clear streams. The meadows were still winter brittle, and banks of snow lingered on the north sides of dry stone walls. The lands around John Burroughs' grave were posted by "C. Burroughs of Grand Gorge, N.Y." Relying upon C. Burroughs' good will, and in search of "what is mine," I ignored the postings and went tramping into the countryside. It was a radiant day. Hawks soared high, on the lookout for small animals lured from hibernation by the warming sun. Woolly-bear caterpillars hunched confidently across the meadows, searching for the perfect place to rearrange their cells into spring's first moths. I saw everything with two pairs of eyes—my own and John Burroughs'. I remembered Burroughs' message: "To know is not all, it is only half. To love is the other half."

Late that night, on a nearby hillside, I scanned the stars for faint lights and saw the Praesepe in Cancer. And I thought of Burroughs, the master of nature's subtle gestures. Burroughs was a confessed creature of the day, but he praised the night. "The gifts of night are less tangible," he said. "The night does not come with fruits and flowers and bread and meat; it comes with stars and stardust, with mystery and nirvana." Now and then night's faint lights revealed themselves to Burroughs, and the heavens opened. His thought went "like a lightning flash" into that abyss, and then the veil was drawn again. Just as well, he said, to have such faint and fleeting revelations of the deep night: "To have it ever present with one in all its naked grandeur would perhaps be more than we could bear."

At the grave site that day in the Catskills I noticed new vines that grew from underneath the marking stones. Young tendrils circulated through the cavities of the enclosing dry stone wall. Green buds reached for the sun in the old eternal way. Spring was casting hints upon the wind. A trait of summer was in the air. On such a day not even the weight of Catskill shale could keep the old man's spirit underground.

·4·

NIGHT CREATURES

W ho goes abroad at night? The glowworm, trailing phosphorescence. The whippoorwill and the cuckoo, self-announcing. The bullfrog and the cricket, contrapuntal, out of tune. The owl and the moth are abroad at night; one is the other magnified. The hedgehog is abroad, with spikes tipped by stars. The woodcock, in flirtatious circles. The slug and the snail, on threads of slime. Into *collied night, sable-vested night*, go specters. And incubi and succubi, with sinister intent. Werewolves transfigure. Vampires seize and suck. Spars of sailing ships are struck with St. Elmo's fire. The *fata morgana* beckons. Poets go abroad at night: "This is the light of the mind, cold and planetary. The trees of the mind are black." And astronomers. When the sun goes down, astronomers rise into their element like shades and badgers.

Tonight I walked a dark road on a high hill in the west of Ireland. The Dipper lifted its overbrimming bowl in the northeast. Seven great stars, seven sages, seven wise men, seven bears, seven bulls. Seven brothers who courted the seven sisters of the Pleiades and stole one of them away (look! there near the star Mizar, the missing sister). I have often been surprised by the way people who do not know the constellations will instinctively and with ease pick out the Big Dipper. I have sometimes wondered if the pattern of those stars is genetically impressed upon the human brain, the way certain birds are born with the ability to recognize and follow the constellations on their migrations. Certainly no group of stars in northern skies has a longer or more prominent history. The Dipper is a repository of human brooding and human dreams.

The Big Dipper is officially Ursa Major, the Great Bear. The constellation has been known as a bear not only in the Western tradition, but also among the Indians of North America. It is difficult to recognize the figure of a bear in the pattern of the seven stars. Some chartmakers try to make a satisfactory outline of a bear by including nine or ten of the fainter surrounding stars. To me, this seems unlikely; the seven bright stars of the Dipper are far more prominent than the included others. The stars of the Dipper, like all stars, slowly change their positions on the celestial sphere (astronomers call this *proper motion*), and it has occurred to many people that the constellation may have become Ursa Major at a time when the pattern of the seven stars more closely resembled the figure of a bear than it does today. But it is not difficult to calculate from present proper motions the positions of the Dipper stars at any time in the past (I have done it myself for many different eras), and at no time within the span of human history does the constellation evoke—to my mind at least—the outline of a bear. Then again, there is the delightful alternative suggestion that Ursa Major received its name at a time in the past when bears more closely resembled dippers!

In ancient China the seven stars of the Dipper were the Celestial Palace of the Immortals. In Ireland they were King David's Chariot; in Scandinavia, the Great Wagon. To the Teutonic tribes they were the Chariot of Thor, and to medieval Christians, the Heavenly Chariot that carried Elijah to Heaven. In English tradition the Dipper is Charles's Wain or the Plough.

Who would surrender the Big Dipper? Who would forgo that wagonload of dreams or chase the Great Bear to a permanent lair? One of Thoreau's companions claimed that a man could get along without the stars, though he would be considerably reduced in his circumstances; they were, he said, a kind of bread and cheese that never failed. The stars are bread and cheese, bread and wine, talismans and covenants. What would poets do without the stars, stars "dropping thick as stones into the twiggy / Picket of trees"? In summer's midnight sun the Eskimos wait for weeks for the Great Bear and find that it has been there all along, hidden in white arctic light, padding in great lopsided circles about the zenith.

❧

Every poet, every fox or badger, every moth or owl who has looked into the night has made one of the most important observations in the history of astronomy: The night sky is dark! The stars shine in a black sky. The Plough that is dragged about the pole furrows black soil. And the darkness of the night sky tells us more about the distant universe than do the stars. Black night is a paradox, rich in meaning. Let me explain.

In 1610 Johannes Kepler received a copy of Galileo's little book *The Starry Messenger*, a book that outlined the Italian scientist's telescopic observations of the sky. Kepler protested against Galileo's argument that the universe was infinite and contained an infinite number of stars. If this were so, wrote Kepler in a letter to Galileo, the entire celestial sphere would blaze with a light as brilliant as the sun. In an infinite universe, no matter which way we looked, our line of sight must eventually terminate on a star, just as a person in a wide forest must in any direction eventually see the trunk of a tree. Since the night sky is manifestly *not* as bright as day, the universe (concluded Kepler) cannot be infinite.

A hundred years later, the English astronomer Edmund Halley proposed that Kepler's argument would not be valid if the light from very distant stars were so weakened by its journey toward us that it could not be detected. But Halley's counterobjection is flawed. The intensity of light from any source weakens in proportion to the square of the distance of the observer from the source, just as Halley proposed. But the volume of space in an infinite universe increases in every direction *also* as the square of the distance, and if stars are distributed randomly in space, then the number of stars emitting light must also increase as the square of the distance. The two effects exactly cancel: The intensity of light diminishes with the distance of the source, and the number of sources increases. So Kepler is apparently correct; in an infinite universe the night sky should be as bright as the sky of day. The light of the seven stars of the Dipper should be drowned in the combined brilliance of the total universe of stars.

Kepler's argument was reasserted by Heinrich Olbers in 1826, and it has come to be known as Olber's Paradox: *If the universe is infinite and uniformly sprinkled with stars, then there should be no night.* There have been many attempts to get around the argument. Some scientists protested that the presence of interstellar gas and dust would absorb and therefore diminish the light of distant stars; but it can be shown that if interstellar gas or dust absorbed starlight, it would eventually become hot enough to reradiate the same energy and therefore maintain the brightness of the sky. The discovery that stars are not uniformly distributed but are instead clumped in galaxies also fails to resolve the paradox; Olbers' argument can be applied to a universe of galaxies as well as to a universe of stars.

But if the universe is *not* infinitely large, then of course there is no paradox. Or if the universe is so young that light from distant stars has not yet had time to reach us, then the paradox is also resolved. The night sky *is* dark, that much is incontrovertible; it would seem we must conclude that the universe is either not infinitely large or not infinitely old. Either conclusion was deeply disturbing to the astronomers of the previous century. A finite universe was impossible to imagine; what was beyond the edge? A beginning for the universe in time seemed to violate the laws of physics—which precluded the creation of something from nothing—and require the special intervention of a deity. The rational turn of mind of nineteenth-century science did not admit of edges or beginnings. The dark night sky put astronomers in a bind.

❧

"Look at the stars! look, look up at the skies!" sings Gerald Manley Hopkins. And I look, and I'm knocked off my feet. "O look at all the fire-folk sitting in the air! The bright boroughs, the quivering citadels there! The dim woods quick with diamond wells; the elf-eyes!" It is more than the human heart can bear. I walk a dark road on a high hill, and the stars of the Dipper burn with the intensity of a welder's arc. The Dipper spins on the pole like a Catherine wheel, now diving toward the horizon, now flying high into the air. First goes yellow Dubhe, *Thahr al Dubb al Akbar*, "the Back of the Great Bear." Then Merak, "the Bear's Loin," lined up with Dubhe on the big hand of the sky's clock; these are the "pointer stars" that swing from the pole like a child on a tether. Next Phad, "the Thigh," and Mergrez, "The Root of the Tail"; faint Mergrez, where the handle meets the bowl. Then out along the handle: first Alioth, then Mizar with its wonderful faint companion Alcor (the stolen sister!), and finally Alkaid. Zeus "flung them through the air, in whirlwinds to the high heavens, and fix'd them there." Zeus, the greatest of the gods, lusted for the mortal Callisto. Beautiful Callisto was a huntress who roamed the mountains of Acadia in search of game. Hera, the wife of Zeus, became jealous of her husband's adulterous obsession and changed the unfortunate Callisto into a bear. As a bear, she cowered in the forest, terrified of both man and beast. One day her son, Arcas, came upon her. Happy Callisto stood up on her hind legs to welcome him! Thinking himself attacked, Arcas readied his bow. But Zeus, looking down from Olympus, saw the terrible event unfolding and with instantaneous magic changed Arcas into a little bear. Then he flung mother and son into the high heavens, fixed forever in the form of bears.

Often I have gone stalking the dark spaces of Ursa Major for galaxies. With a moderate-sized telescope, two of the nearest and brightest galaxies can be found in that constellation. M81 and M82 were discovered by Bode at Berlin in 1774, two fuzzy patches in the night sky. The comethunter Messier added them to his catalogue of fuzzy spots in 1781, and ever since we have known them by their Messier numbers. M81 is the more splendid object, resolved in observatory photographs into a dazzling pinwheel of 200 billion suns, but reduced to a misty blur in the eyepiece of the amateur's telescope; if M81 were at the distance of the seven Dipper stars it would fill our sky with a terrible light, wheeling and glittering, suns as thick as stones, bread and cheese in superabundance. Its sister galaxy, M82, is spindle-shaped, enigmatic, apparently wracked with violent convulsions, stars gone haywire, a galaxy regurgitating its own mass. The core of M82 has been shattered by a titanic explosion involving the mass of a million suns, stars flung like drops of water from a shaking dog, planets shaken from stars, green worlds and blue worlds blown to kingdom come. *Spare us, O Lord, thy wrath.* This is the uncaught arrow of Arcas, plunged into his mother's heart.

Five of the seven stars of the Big Dipper are part of a true cluster, about 80 light-years away. Dubhe lies just beyond the cluster at 105 light-years. The blue-white giant Alkaid is far beyond all the others, double the distance of Dubhe, at 210 light-years. The Dipper stars are all stars of our own Milky Way Galaxy, and relatively close. M81 and M82 are 7 million light-years away; they are other "island universes" far out in the sea of intergalactic space. The explosion that tore M82 apart took place 8½ million years ago—Earth time—1½ million years before we see the galaxy now. Into the shock wave of that cataclysm I go from this high hill, from this dark night. *Look, look at the stars, look up at the skies!* Across light-years. Galaxies burning in a black sky. This is the light of the mind.

❧

The resolution of Olbers' Paradox waited until the twentieth century, and then it came from an unexpected quarter. Between 1912 and 1928, the astronomer Vesto Slipher succeeded in obtaining spectra of the light from forty galaxies. At that time, the so-called "spiral nebulas" were not yet recognized as distant star systems, the equals of the Milky Way. Slipher passed the light of the galaxies through a prism and examined the rainbows of their colors. The light of the galaxies was typical starlight, but the wavelengths of the light were slightly lengthened—shifted toward the red end of the color spectrum. The red shift of the galactic light implied to Slipher that the galaxies were moving away from us. This is called the Doppler effect, and it is the same effect as the sudden drop in pitch one hears when a truck roars past on the highway. As the truck approaches, the wavelengths of its sound are compressed, and the pitch is heightened. As the truck draws away, the wavelengths are stretched out and the pitch is lowered. The same applies to light waves. The light of sources that are moving toward us is compressed and shifted toward the blue—or short wavelength—end of the spectrum. The light of receding sources is stretched and reddened. M81 is rushing away from us at a velocity of 48 miles per second. M82 is receding at 240 miles per second.

At first it might seem strange that the galaxies are flying from us. Are we such unattractive company? Are we at the center of a cosmic repulsion? The answer is more straightforward and was soon forthcoming. While Slipher was obtaining the spectra of galaxies, Edwin Hubble and Milton Humason at the Mount Wilson Observatory were successfully measuring the distance to the galaxies. They discovered an unexpected relationship: Not only are the galaxies receding from us, but the velocities of recession are directly proportional to the distances of the galaxies. This is precisely the relationship one would expect if the texture of the universe were uniformly expanding—everywhere. The galaxies, it seems, are rushing apart. Space is inflating. Our galaxy is not at the center of this cosmic repulsion; residents of every other galaxy would also see their neighbors receding.

If the galaxies are rushing away from one another, it

is not hard to calculate from their relative velocities and present separations that they must have been together 15 billion years ago. The outrush of the galaxies began in a universe-embracing explosion. The primeval fireball. The Big Bang!

Here, then, is the resolution of Olbers' Paradox: *The night is dark because the universe is expanding.* The universe may or may not be infinite in extent. But, in any case, it is finite in time. It had a beginning 15 billion years ago. There is no way we could receive the light from stars more than 15 billion light-years away, for no such stars existed. Furthermore, the light from receding sources is weakened by the outrush. Distant galaxies, if they are moving away from us, do not contribute their share of light to the brightness of the night sky as required by Olbers' Paradox. The universe is too young for the night sky to blaze with light!

❦

Our lives are rounded with a dark. "How insupportable would be the days, if the night with its dews and darkness did not come to restore the drooping world," wrote Thoreau. "As the shades begin to gather . . . we steal forth . . . like the inhabitants of the jungle, in search of those silent and brooding thoughts which are the natural prey of intellect." The night sky is the hunting ground of the mystic and the philosopher, the scientist and the theologian. I have walked along a dark road on the brow of the hill for an hour. The Great Bear has clumped one-sixth of the way up the slope of the eastern sky. The Little Bear is swung by his tail about the pole. Zeus lusts for Callisto. Callisto lays down her quiver. Hera spins jealous designs. Callisto loves Arcas. Arcas is afraid. The arrow is poised on the bow.

"Tonight, in the infinitesimal light of the stars / The trees and flowers have been strewing their cool colors"; this is the third time in this night of meditation I have quoted from poems of Sylvia Plath. Each day is a little life, and each life is rounded with a little dark. The galaxies are rushing away from us, diluting their brilliance, darkening the sky. *Night is the universe's youth.* There are hedges at both sides of the road along which I walk, hedges of bramble and honeysuckle and fuchsia. The lanterns of the fuchsia dim their pale light, all their gaudy puce and scarlet gone, flown away with the galaxies. The universe is young. I walk in its adolescent light. Now there is time, "the trees may touch me for once, and the flowers have time for me."

5

BEGINNINGS

T he spring comes slowly up this way," sang the poet. And this morning it was there, in a meadow beaten flat by winter, hiding in a cavity of grass abandoned for a deeper burrow by some still-sleeping creature. Spring was there in that borrowed nest. It was the first day of April, and the meadowlark was back!

I didn't see him. You seldom see an early meadowlark unless you come close enough to scare him up from his hiding place. But his long, slurred, double-noted call is as sure a sign of spring as a crocus and, like a crocus, a trifle premature. In a month or two the meadow will be lush and green, but today the thought of spring is tinged with heartache. Like the song of the meadowlark.

There's no mistaking a meadowlark's call. I can whistle it. But how do I describe it? A dive from a springboard into icy water? An April wind breaking on window glass? No, neither is quite right. I go to my bird books. The *Golden Field Guide* provides a very scientific "Sonagram" of frequency versus time. The song, I can see from the graph, is about two seconds long and ranges from three to four octaves above middle C. That is of no use at all. Peterson's *Guide* is a little better: "Two clear, slurred whistles, musical and pulled out." *Tee-yah, tee-yair*, tries Peterson, striving for objectivity, and that gets us close to the sound but not to the strange, sad music.

As usual, one has to go back to the older guidebooks for something closer to the reality. Chapman's classic *Handbook of the Birds*, published in 1895, catches a bit of it: "The meadowlark's song is a clear, plaintive whistle of unusual sweetness." Ah, that's better—the sweet and the sad. But in this matter, as in all things pertaining to birdsong, F. Schuyler Matthews' seventy-five-year-old *Field Book of Wild Birds and Their Music* does it best. The song, says Matthews, is "unquestionably pathetic, if not mournful." And with his characteristic extravagance, Matthews transcribes the meadowlark's call as the first two bars of Alfredo's song in *La Traviata,* but sung (of course) the way Violetta sings it when she discovers she must give up Alfredo. The sweet and the sad. The song of the meadowlark. Spring!

Why are beginnings touched with sadness? The birth of a child, the beginning of a new year, the first triumphant notes of a Beethoven symphony, the call of a wild bird in an awakening meadow—all moments of promise, of joy even, all infected with a strange, sweet melancholy. Does the meadowlark know something I don't know? That acolyte in black-and-gold vestments has a secret. From his hiding place in the crumpled grass he lectures on existential philosophy and discourses on roses and thorns. His hopeful announcement of spring is *sicklied o'er with the pale cast of thought.* Beginnings wear their endings like dark shadows.

🍏

A creation myth from the ancient Mediterranean (translated into English by Charles Doria and Harris Lenowitz) has God bring all things into being with seven laughs. *Hha Hha Hha Hha Hha Hha Hha.* Does God know what the meadowlark knows? Did He roar that first *Hha* with a twinkle in His eye? Was it all a joke? A prank? Today is April Fool's Day. A perfect day for the beginning of spring. A perfect day for the Creation.

Here is how Doria and Lenowitz translate the first laugh: *Light (Flash) / showed up / All splitter / born universe god / fire god.* Those lines are 2000 years old, translated more or less literally from the ancient texts. Still, it would be hard to find a more apt account of the modern scientific view of Creation. The "Big Bang," the astronomer Fred Hoyle called it contemptuously, and the name caught on. A better name might be the "Big Flash," the *all-splitter.* Fifteen billion years ago there was nothing. Then God laughed. An infinitely dense and infinitely hot seed of energy sprang into being from nothingness and flowed instantly into matter. According to current cosmological thought, that first laugh took a billionth of a billionth of a billionth of a second, and when it was all over, the universe was off and running.

Elementary particle physicists and cosmologists sit down with their equations and their lead pencils and their yellow pads and they reconstruct theoretically the first microseconds of the Creation. This is hubris on a cosmic scale, and we'd be fools not to take them seriously. These are the people who detected quarks and quasars, who turned matter into energy, who charted the trajectories of particles that live for a millionth of a millionth of a second and leave a trace on existence more subtle than the scrape of a fingernail on glass. It is only natural that such people should ask: When and where did it all begin? And how? And finding the answers is just a matter of working backwards, reversing on paper the outrush of the galaxies, mathematically sticking the toothpaste back into the tube.

The beginning point for recent speculation about the origin of the universe was the discovery of the recession of the galaxies. The universe is expanding! The texture of space is blowing up like a balloon, swelling like a loaf in the pan. And like dots painted on the surface of the inflating balloon or raisins in the rising loaf, the galaxies are carried away from one another as space expands. Surprisingly, this very behavior for the universe had been predicted early in the century by Einstein, as he fiddled with his equations of general relativity. The equations seemed to insist that the universe would swell like leavening dough. The result was so bizarre and unexpected that Einstein refused to accept it. He added a gratuitous constant to his equations to repress the inflation and in the process, of course, adulterated the simple elegance of the mathematics. When Edwin Hubble later announced that the universe did indeed seem to be puffing up, Einstein rushed to Mount Wilson to have a look through the big telescope. (While on the mountain, someone remarked to Einstein's wife, Elsa, that the giant instrument with the 100-inch mirror was used for determining the structure of the universe. "Well, well," responded Elsa, "my husband does that on the back of an old envelope.") Einstein was impressed with what he saw at Mount Wilson and struck the offending constant from his equations. That constant was, he said, the biggest mistake of his life.

If the galaxies are flying apart, then they must have once been together. If we run the movie in reverse, the galaxies fly toward one another, accelerating. They wring out the vast empty spaces. They crush their stars together like wet sand in a fist. Stars are squeezed into stars, matter into matter. The density of the universe soars. The movie ends in a blinding flash of pure energy, infinite, singular, *born universe god / fire god*, the Beginning.

Contemporary cosmologists can calculate the conditions of the universe at every instant of its history from well-known laws of physics. Relying upon those calculations, we now run the movie forward. The universe begins 15 billion years ago in a blinding flash, the primeval fireball, from an infinitely dense seed of pure energy. The seed is not "somewhere," it is everywhere. Space itself is created as the universe begins to expand. A ten-millionth of a trillionth of a trillionth of a trillionth of a second later, elementary particles—quarks and electrons—flicker in and out of existence against a background of radiation, dissolving and reappearing, dissolving and reappearing, fragments of the *all-splitter*, Creation struggling to be born.

A millionth of a second after the Beginning, the quarks join in a dance of threes to form protons and neutrons. Another thousandth of a second and protons and neutrons stick together to create the nuclei of the light elements. Matter and antimatter annihilate each other in an orgy of self-destruction. A flood of neutrinos goes flying into the future.

Time passes, the universe cools. Atoms form. Then

galaxies. Then stars. Quasars blaze like luminous beacons at the cores of galaxies. Space expands. A few billion years after God's first *Hha* the universe begins to look like home, although it will be another 8 billion years before the solar system condenses from galactic cobwebs in a dusty corner of the Milky Way. By the time the Creator gets to his fifth laugh (in the myth translated by Doria and Lenowitz) he *smiles weeps*—the sweet-sad song of the meadowlark all over again—Creation's only begun and already winter's just around the corner.

❦

Not long ago physicists and astronomers despaired of ever knowing what came before the instant of the Big Flash. At that singular moment in the universe's history, their equations shot off to infinity like skyrockets. The numbers for the density and temperature of the universe increased without limit, mathematical mountains that could not be climbed or seen over. Space and time collapsed into a dread singularity, a numerical bottomless pit that became increasingly narrow and increasingly deep until it was a thread of calculation too long and too thin to follow. There is a street in my town like that. In the words of a local historian, the street becomes an unpaved road, then a track, then a path, then a squirrel trail that runs up a tree. Tracing the universe mathematically back to the Beginning is like following that street until you find yourself up a tree with no place to go. The question of what *caused* the "Big Bang" was considered intractable, if not meaningless. Creation out of nothing seemed to violate the law of conservation of matter and energy, but physicists shrugged and said that that was the way it was and there was nothing more they could say about it. A laugh or a day's work on the part of a Creator was as good an explanation as any other.

Lately, a new generation of young cosmologists have

become somewhat bolder in their speculations and have begun to squeeze from their equations a glimpse of the world beyond the Beginning. They draw upon recent discoveries about the behavior of matter at high temperatures. They twist their theories of space and time into new designs that will encompass what we have learned about quarks and neutrinos, the elusive W and Z_0 particles, and the other subatomic building blocks of the universe. And out of this comes an astonishing prediction: Our universe, the universe that began 15 billion years ago in a fireball of radiant energy and matter, may be just one universe among many. Universes may boil like bubbles from a greater matrix of hyperspace and hypertime, popping into existence as quantum fluctuations in that superspace with the positive energy of stars and galaxies balanced by a negative potential energy of gravity, a bubble of Creation that adds up to zero, a bubble that exploded from nothingness without violation of the laws of physics, a bubble that contains our space and our time and our Milky Way and its billions and billions of sibling galaxies. If the conjurers who do these calculations are right, then universes are popping into existence all the time, and our starry night is the interior of a single bubble of galaxies in a frothy ongoing spontaneous creation.

If this stuff makes your head spin you are in good company. Having a head that can spin is a requirement of all great science. Copernicus' head was spinning when he recognized that the Earth was one planet among many. Newton's head was spinning when he told us that the sun was just another star. Hubble's head was spinning when he proved that the spiral nebulas were Milky Ways. Alan MacRobert writes: "Whatever else is in the character of nature, as we see it unfolding into ever more abundant vistas before our patient inquiry, we find that it does not economize on its size and richness." God's *Hha Hha Hha* was no snicker, but a roaring belly laugh.

If the meadowlark's sweet-sad song harkens our thoughts back to the Beginning, to the start of that sweet-sad season that is the universe, then the red-winged blackbird's raspy call is the voice of the hidden superspace, the nothingness which is something, the ground of random fluctuations beyond and before the Beginning. Around my part of New England, the red-winged blackbird arrives first—before spring, before the skunk cabbages, before the pussy willows, before the fiddlehead ferns, before the meadowlark. And this year the redwings were earlier than ever. It was the second week of February when they arrived. I heard them croaking in the high oaks by the pond, pretending to be sociable. Soon they would string themselves out in smaller trees along the brook, each defending with a raucous belligerence his own private stretch of water. And the crows reacted! All winter long the crows had had the place to themselves; now they cawed and shrieked and wheeled and dived and otherwise protested the invasion by the upstart blackbirds. The imperious redwings, perched in their treetops, ignored it all.

The red-winged blackbird is an aloof bird, an arrogant bird. He is saved from a kind of sleek ugliness only by the flaps of color on his wings. His raspy voice could saw wood. The redwing is the voice in the wilderness, clothed in camel's hair, the eater of locusts and honey. We forgive him everything. We forgive him because he is the beginning of the beginning, the beginning of the beginning of the beginning. He gets here first, sometimes before the snows have gone, smelling of sweet-sad spring. From his mouth bubble universes, beyond sweetness, beyond sadness, beyond the peaks and valleys of the physicist's equations. This is the Beginning. Upon the seventh laugh (I still follow Doria and Lenowitz) *GOD looked / HE went pepepepepepepepepepepepepepepep / just like a bird.*

6

AN ANCIENT
BRILLIANCE

I n the beginning there was light.
In the first moments of Creation, subatomic particles flickered in and out of existence against a background of radiation. The universe was a seething fireball of matter and energy. One second after the beginning, the temperature of the universe had dropped to a chilly 10 billion degrees, and the creation of matter ceased. Protons, electrons, and neutrons danced upon a sea of light. Still the universe was filled with blinding radiation.

When a million years had passed, the universe had cooled to the point where charged particles could hold together against the pressure of radiant energy, against the flash of the new, against the *all-splitter*. Electrons linked with nuclei to form atoms. The hard stuff of the universe was born, mostly in the form of hydrogen and helium. This was light stuff, to be sure, the stuff that floats balloons, but now the universe had a foundation, a bedrock of atomic matter on which to build, bone and nail. The reign of light subsided. The universe became transparent. But there was still to be another chapter to the ancient brilliance. Sometime within the next billion years the galaxies began to form—and not long after, so did the quasars.

❦

Today in the mail I received a poem that a friend had snipped from a journal. It describes the supposed effects of a Catholic education at the hands of priests. The poem ends with these Joycean lines: "They flushed sin from the coverts of our souls with / fear and drove God's sacred plover crying into the upland rain / where it remains." To the poem my friend affixed a note: "Is this what happened to your plover too?" Well, I don't know what happened to my plover, but it has certainly flown the coop.

Most plovers are shore birds, sandpiperish birds. But the upland plover makes its home on high heaths and moors, on hillside fields and rough meadows. Its voice is like the whistling of the wind and can be heard even at night. The upland plover is a shy bird. It is the color of dry grass. In the rare event that one is flushed, it takes to air with a soft, bubbling whistle. "During their migrations," says one of my handbooks, "one may clearly hear these sweet notes from birds traveling beyond the limits of human vision." If the poet wanted an image for the absconded God, he could have found none better than the upland plover.

I can't say exactly when it was that the God of my youth took to the upland rains. He was not *driven* from my soul. His flight was no fault of my teachers'. My lapse from faith occurred not long after my graduation from college, at the end of a period of intense belief during which His face seemed palpably near. Those were heady times to be a college student and a Catholic. We read the French Catholic authors: Bernanos, Bloy, Péguy, Mauriac, Maritain, and Teilhard de Chardin. We read the English authors: Chesterton, Greene, Waugh, Hopkins. We read Sigrid Undset and Antoine de Saint-Exupéry. And sacred plovers leapt from every page, took to wing in coveys, and made a tumult with their wings that drowned the thin voice of doubt. Emily Dickinson called hope "the thing with feathers." The plover was our hope. The plover was Hope, Faith, and Charity.

Then one day I woke up and the plover was gone. The mockingbird on the rooftop was making beautiful music, but the plover had vanished into the uplands. I turned to my science books and got on with the business of life. But something was missing—"the thing with feathers." In God's absence I have tried to make a sort of theology of ornithology. I listen for the crisp double note of the chickadee in winter, and the sweet-sad song of the meadowlark in spring. I wait for the veery's spiraling aria at midday, and the killdeer's razor-sharp call at dusk. But the sacred plover continues to reside in the uplands with the wind and the rain, and something deep inside me knows that it is gone forever.

In the dark hours of the night, in starlight, I listen for

the scrawny cry. Is it the wind or the plover there on the hill behind the house? The plover is invisible because its coat is the color of dry grass and its voice is the voice of the wind. The nineteenth-century ornithologist Frank Chapman wrote of the upland plover: "One may ride over a prairie upon which, at first glance, not a plover is visible, and find, after careful scrutiny, that dozens of birds are scattered about." *And so does God hide in dry grass and rain and night's faintest lights.*

❦

The quasar 3C 273 is available to the backyard telescope. I've seen it in my fourteen-inch scope. I would never have known what it was that I was looking at if I hadn't had a finder chart in front of me. Quasar 3C 273 looks just like any faint-bluish star, like any star of the thirteenth magnitude, a star on the very limit of vision for the telescope I was using. Quasar 3C 273 looked just like any one of the thousands of faint thirteenth-magnitude stars in the constellation Virgo. To find 3C 273, I set my telescope on the bright star Porrima and then scanned a few degrees to the northwest until the stars that matched the finder chart came into the field of view. Porrima is thirty-five light-years away, close by cosmic standards, in our own neighborhood of the Milky Way Galaxy. Only about two or three hundred stars are closer to us than Porrima, most of them so small and faint as to be invisible even in my fourteen-inch scope. Quasar 3C 273 is 1½ *billion* light-years away, a hundred million times more distant than Porrima, beyond all of the stars of the Milky Way Galaxy, beyond billions of other visible galaxies. Quasar 3C 273 is 1½ billion light-years away and 1½ billion years back in time. That unspectacular point of light in my telescope was as close as I'll ever get to witnessing the birth of the universe. That point of blue light was my faint glimpse of the blinding radiance of the Creation.

Nature provides us with a way to see back to the early days of the universe. It takes time for light from distant sources to reach the Earth. When astronomers look at distant objects through their telescopes, they are looking backward into time. The stars are light-years away. The galaxies are millions of light-years distant. The quasars are removed from us by billions of light-years, and therefore by billions of years of time. The quasars are the most distant objects yet observed. The light of 3C 273 is the most ancient light that will ever enter my eye.

❦

Quasars were discovered in 1963 by Maarten Schmidt of the California Institute of Technology. Schmidt caught on photographic plates the light spectra of a few starlike objects that had attracted the attention of astronomers by their copious emission of radio energy. The objects were dubbed "quasi-stellar radio sources," or "quasars" for short. The spectra of these objects were unfamiliar; they were unlike the spectra of any other stars. Nor did the colors of the spectra match the radiation of known forms of matter. Astronomers were baffled. Then, with sudden insight, Schmidt recognized in the enigmatic spectra the disguised features of the spectrum of hydrogen; *the pattern of wavelengths typical of hydrogen radiation had been shifted drastically toward the red (or long-wavelength) end of the spectrum!* The lengthening of the wavelengths could only be the Doppler effect, the stretching out of the wavelengths of the light because of a motion of separation between source and observer. If *all* of these objects are receding from us, it must be because they share in the general expansion of the universe. The unusually large shift of the spectrum signified to Schmidt that the objects are very far away, more distant than anything previously observed.

In a uniformly expanding universe, the amount of reddening of the light of distant sources is directly proportional to the distance of the source from the observer. 3C 273 was one of the radio sources considered by Schmidt. The red shift of 3C 273 corresponded to a velocity of recession of 30,000 miles per second, or 16 percent of the velocity of light. 3C 273 and the other quasars in Schmidt's sample were apparently billions of light-years away. They were farther away than the most distant visible galaxies. Their light had been traveling toward us since early in the history of the universe.

But the quasars shine far more brilliantly than the distant galaxies, and that is why I can see one burning there on the marches of the universe even with my fourteen-inch scope. What could these objects be, beckoning across the eons and tantalizing us with a mysterious glimpse of an earlier era? From their flickering light, astronomers could deduce that the quasars are small, perhaps no larger than our solar system. And yet they are a thousand times more luminous than entire galaxies, brighter than a hundred billion suns.

Some astronomers are reluctant to believe that anything so small can be intrinsically so bright. They hope to show that the quasars are relatively nearby, in which case there is no need to believe that they are so luminous. But if quasars are nearby and do not share in the expansion of the universe, then what causes the reddening of their light?

The cosmological distances of the quasars are now almost universally accepted. But no one yet knows the true nature of these strange objects, these enigmatic travelers from the epoch of Creation. There is a growing consensus that quasars are the bright nuclei of very distant galaxies. Perhaps they are chains of supernovas detonating in the star-rich central regions of galaxies that are too far away to be otherwise visible. More likely, they signal the infall of matter into central black holes, massive centers of gravitational attraction at the cores of young galaxies, bottomless gravitational pits. Plunging to oblivion, the infalling matter sheds the enormous amounts of energy that power the quasars.

Surveys have shown that the number of quasars in the sky increases with distance. Apparently these objects were more common in the early universe than today. If the black-hole model of the quasars is correct, then a typical stage in the evolution of young galaxies was the formation of massive central black holes. Stars fell toward these colossal cosmic sinks, whirling into them like water circling a drain, accelerating to nearly the speed of light, shedding abundant energy in the form of radiation of all wavelengths. Millions upon millions of stars fell to oblivion into the open mouths of the galactic black holes. These spasms of violence that marked the creation of the galaxies have now mostly subsided, and galaxies—our Milky Way included—have settled down to a more tranquil existence.

Not only were there more quasars in the ancient universe, but they were also more luminous. We can envy the brilliance of those early skies. The galaxies were closer together than they are now, and they burned with the light of many hot blue stars. At the centers of those great wheels of light, streams of matter plunged into black holes, pulled by gravity into knots of incredible density and permanent blackness—stars, planets, moons, rain and wind, all absconded, gone. As the matter fell, it gave off energy that caused the galactic nuclei to glow with a light greater than that of all the stars in the heavens. The universe blazed with those luminous beacons. It was a time of light.

❦

On a single photograph of the night sky made with a large wide-field telescope, there may be as many as 200,000 images that appear to be stars. A few hundred of these will turn out to be quasars. These monstrous brilliances from the past hide among the stars like plovers in dry grass. Their low-pitched hydrogen radiation is lost in starlight like the plover's song on the wind. If there had been humans on the Earth 12 billion years ago—if there had been an Earth!—they would have seen quasars burning on every side in the night sky with the intensity of our brightest stars. And what of the core of our own Milky Way Galaxy? Was it too, in those days, a quasar? If so, its light would have bowled us over, shaken reeds, bent grass, fiercer than a thousand suns, jasper-stone, pure crystal. The stars then would have been hidden in its light, the sweet influence of the Pleiades bound in brightness, the bands of Orion loosed.

Paul was thrown to the ground by a light that struck with steel-furnace force. John saw it burning with the intensity of the midsummer sun. Teresa of León was bathed all over with a resplendent glow. The mystics, it seems, directly experienced the radiance of Creation. Gone, all gone now, that ancient brilliance, that flood of pure light that billions of years ago held the young universe in thrall. When I decided to go looking for 3C 273, a relatively nearby quasar and the brightest in our sky, I waited a week for a moonless night and then another week for clear skies. I wanted Virgo as high as possible above the horizon, which meant rising before dawn on a cold February night. A wait of a few weeks seemed not unreasonable; the faint light I sought had been traveling toward me for more than a billion years, even as the Earth and the quasar raced apart. When the light that entered my telescope left 3C 273, there was no life on Earth other than single-celled microscopic creatures afloat in the sea. Those creatures were blind to the gorgeous young Milky Way, to the myriad bright blue stars that ornamented their eyeless night. I swung my telescope along the Dipper, making "an arc to Arcturus," then on until I "spied Spica." Spica is the brightest star in Virgo; it was my beacon to the quasar. I

flew the telescope by the seat of my pants (rather than using the setting circles), moving cautiously from Spica to
Porrima, then into darkness, following the guide chart
across an archipelago of faint stars. At last 3C 273 tiptoed
into my field of view, circumspect and anonymous across
the light-years, blinding Creation compressed to a point of
light, as in a *camera obscura*. It was like the cry of a bird
in the upland rains, faint and far away. As I watched the
quasar, the sky began to brighten, and almost immediately
the quasar was lost in the sun's glare. I thought of the last
words of *Walden:* "The light which puts out our eyes is
darkness to us. Only that day dawns to which we are
awake. There is more day to dawn."

7

SNAKES AND
LADDERS

The quasars, the most distant visible objects in the universe, show in their spectrum the familiar light of hydrogen. In the dust and gas clouds that bank the stars of Orion, astronomers have detected the telltale radiation of carbon, oxygen, nitrogen, sulfur, and silicon. In the rarified environment of interstellar space they have found water, ammonia, acetylene, alcohol, and a host of other common terrestrial molecules. The universe, it seems, is consistent in the stuff of its construction. Galaxies and stars, planets and moons, bacteria and blue whales, they are all merely arrangements of ninety-two atomic elements.

Give me the ninety-two elements and I'll give you a universe. Ubiquitous hydrogen. Standoffish helium. Spooky boron. No-nonsense carbon. Promiscuous oxygen. Faithful iron. Mysterious phosphorous. Exotic xenon. Brash tin. Slippery mercury. Heavy-footed lead. Imagine, if you will, a chemical storeroom stocked with the ninety-two elements. Pop the corks, open the valves, tip over the boxes and canisters. Watch what happens. What a to-do! Energy released and absorbed. Atoms linking valancies to make molecules. Simple molecules reassembling their parts to make complex molecules. Sparks, flames, and flashes of light, a commotion of combination and alliance. These are elements with a rage to order. These are elements with a zest for life.

There is more to the elements than meets the eye. Rubidium is silver; its salts turn flame to crimson. Calcium regulates the heartbeat; there are two pounds of the stuff in my body, mostly in the teeth and bones. I suck in a bucketful of nitrogen with every breath; it is the backbone of proteins and TNT. Arsenic will kill you; some of its compounds are medicines. Inert argon makes itself useful by doing nothing. Take an atom of inert gassy argon, toss in an extra proton and an extra electron, and you have solid explosive potassium; the natural radioactivity of potassium in the Earth's crust helps cause the mutations that drive evolution.

Why have philosophers given this magical, versatile stuff such short shrift? Since early in the history of Western thought, matter has been assigned the lowliest role in the drama of creation. All of the big parts went to light, force, energy, spirit. Matter was the dross of the universe, the chaff, the bottom link of the chain of being, the lowest rung on a ladder of value that reached from the ponderous center of the Earth to the highest heaven. Matter was Caliban to light's Ariel. Matter was spirit's burden.

Twentieth-century physics has gone a long way toward rescuing matter from the philosophical doldrums. Recent investigations in high-energy particle physics suggest that the great chain of being is more like a snake than a ladder, a snake that bites its own tail. Probing inside the atom, physicists have encountered the universe of the galaxies. At the heart of matter they have glimpsed the radiance of the Big Bang. Caliban has taken off his mask, and he is Ariel.

❦

I have a friend named Anne Saxon who is descended from a long line of stonecutters. Her brother Ian still cuts stone—he maintains the family's monument business. He cuts some of the finest headstones in New England. Stone is Ian Saxon's life. *Saxon* means "stone."

Anne is a scholar. She lives in the world of ideas, of art, of light. "The light enters my eye oddly," she maintains. She moves the furniture, she wallpapers, she paints. She is shaping light, she says, trying to achieve a certain airiness in matter. She installs skylights. She knocks out walls and puts in windows. At Christmas she puts candles in all the windows. She has arranged her life so that she rises before the sun and sleeps soon after the sun sets. She pursues what Emily Dickinson called "that strange slant of light" typical of New England. She is trying to live out "some atmospheric mood," she says; "matter is only there to fill an aesthetic purpose." Anne told me once what she liked about churches; the liturgies of the church, she said, are like the particular radiances of New England. I have a feeling that Anne Saxon goes to church for the same reason she knocks holes in her roof; she is arranging matter for spiritual effect. Isn't that, after all, the meaning of a sacrament?

Ian Saxon is a simple man, in the country sense of the word. He is the natural engineer, the husbandman. Ian has an affinity for the soil. Plants may depend on sunlight for photosynthesis, but Ian takes care of their material needs. He has an unerring instinct for how much fertilizer and how much water a plant needs in order to put down sturdy roots. Ian worries about how things are put together and how they work. For a long time he worried about the discoloration on the family's silver flatware. One day there came in the mail a letter from the local authorities saying that the town water supply had a high salt content and, consequently, that people with a low tolerance for sodium shouldn't drink it. Ian made the connection, and now the silver shines.

Ian has no interest in shifting patterns of light. He is interested in things that endure, things that have a solidity about them—like stone. Granite, says Ian, should be there forever, and he means it. At the monument shop he passes his hands across a new consignment of stone. How will it weather? Will it discolor? Will it streak? Is it soft? Is it hard? He is never satisfied with his work. He has seen the work of the great stonecutters of the past and knows that he is not one of them. But he knows stone. When it became fashionable some years ago to make grave markers of black granite, Ian was bewildered. He knew that black granite was not right for New England. It wasn't native; it didn't work well here. Black granite came from Africa. It was a tropical stone. And it did not letter well. People wanted Ian to use black granite and paint white into the letters. "Paint granite!" he scoffed. "Paint is for preserving wood!" People have asked Ian to put canoes and hunting dogs and horses on headstones. He doesn't like that. He would prefer to carve crucifixes and Stars of David. Stone is forever, he says, and shouldn't enshrine things that are merely silly. He particularly likes to work in Hebrew. He likes the letters, the mysteries of the strange godly language. He goes to the synagogue to make sure he gets it right, and he cuts it into stone and that makes sense. Ian has a New Englander's hostility to waste. Granite, he believes, shouldn't be wasted.

Anne lives in qualities of light. Ian carries with him a dusting of stone. The sun never enters Ian's work shed. He breathes stone. It coats his lungs. It is entrenched beneath his fingernails. Anne rises early to catch the clean light of slanted dawn. I know these two people. They are two sides of the same coin. They are the snake that bites its tail. In Ian's stone dust, there is the light of the angels; in the play of light on Anne's walls, matter dances.

❦

The classical idea of matter was something with solidity and mass, like wet stone dust pressed in a fist. If matter was composed of atoms, then the atoms too must have solidity and mass. At the beginning of the twentieth century the atom was imagined as a tiny billiard ball or a granite pebble writ small. Then, in the physics of Niels Bohr, the miniature billiard ball became something akin to a musical instrument, a finely tuned Stradivarius 10 billion times smaller than the real thing. With the advent of quantum mechanics, the musical instrument gave way to pure music. On the atomic scale, the solidity and mass of matter dissolved into something light and airy. Suddenly physicists were describing atoms in the vocabulary of the

composer—"resonance," "frequency," "harmony," "scale."
Atomic electrons sang in choirs like seraphim, cherubim,
thrones, and dominions. Classical distinctions between
matter and light became muddled. In the new physics,
light bounced about like particles, and matter undulated in
waves like light.

In recent decades, physicists have uncovered elegant
subatomic structures in the music of matter. They use a
strange new language to describe the subatomic world:
*quark, squark, gluon, gauge, technicolor, flavor, strange-
ness, charm*. There are *up* quarks and *down* quarks, *top*
quarks and *bottom* quarks. There are particles with *truth*
and *antitruth*, and there are particles with *naked beauty*.
The simplest of the constituents of ordinary matter—the
proton, for instance—has taken on the character of a Bach
fugue, a four-part counterpoint of matter, energy, space,
and time. At matter's heart there are arpeggios, chromat-
ics, syncopation. On the lowest rung of the chain of being,
Creation dances.

Remarkable progress has been made toward unifying
our understanding of the fundamental forces that hold atoms and the universe together. The Weinberg–Salam theory has succeeded in bringing into a single conceptual
scheme the electromagnetic force that binds electrons to
the atomic nucleus and the "weak" force that governs certain kinds of radioactive decay. The theory called quantum
chromodynamics has enjoyed enormous success in ex-
plaining the "strong" force that girds protons and neutrons
in atomic nuclei. New, highly speculative "super symme-
try" theories promise to unite the electromagnetic-weak
forces and the strong force, and to bring gravity into the
fold as well. Gravity is the force that holds the galaxies in
thrall. This "grand unification" would realize Einstein's
dream of a single all-embracing theory of nature.

Already, the astronomers and the particle physicists
are engaged in a vigorous dialogue. The astronomers are

prepared to recognize that the large-scale structure of the universe may have been determined by subtle interactions of particles in the first moments of the Big Bang. And the particle physicists are hoping to find confirmation of their theories of subatomic structure in the astronomers' observations of deep space and time. The snake has bitten its tail and won't let go.

Until now, the primary instrument for the investigation of matter has been the particle accelerator. These machines are all descendants of the five-inch-diameter cyclotron built by Ernest Lawrence in 1932. The machines use electric fields to accelerate particle beams to high energies, and magnetic fields to guide and focus the beams. The beams of accelerated particles are slammed into other particles at rest or are made to collide with beams moving in the opposite direction. In the debris of these collisions, physicists search for the ultimate constituents of matter. By a strange irony, more and more energetic machines are required to probe matter at ever more delicate levels of structure.

In this country, there are major research instruments at the Stanford Linear Accelerator Center in California, the Fermi National Accelerator Center near Chicago, and the Brookhaven National Laboratory on Long Island. Currently, the most energetic of these machines is the superconducting proton synchrotron at the Fermi lab. That machine is designed to accelerate protons to one trillion electron-volts in a ring of superconducting magnets three and a half miles in diameter. An electron-volt is the energy that can be imparted to an electron or proton by a single flashlight battery. The Fermi lab machine will apply the equivalent power of a trillion flashlight batteries to single protons and knock them into pieces. Then the physicist will look for meaning in the shards.

As the American physicists tuned up their big machines, Europeans raced ahead to win the prizes. The colliding beam accelerator at the European Laboratory of Particle Physics (CERN) near Geneva was the first operational machine to reach energies of several hundred billion electron-volts. In 1983 CERN announced the discovery of the elusive W and Z_0 particles, which had been predicted by the theory that unifies the electromagnetic and weak forces. The discovery provided a stunning verification of the theory (which predicted the particles' existence) and gave a considerable boost to the morale and prestige of European particle physicists.

The discovery of the W and Z_0 particles at Geneva spurred American physicists in their efforts to build the biggest machine yet. The Department of Energy's High Energy Physics Advisory Panel recommended that American physicists abandon work on Brookhaven's half-finished colliding-beam accelerator and proceed immediately with a twelve-year effort to build a multibillion-dollar "superconducting supercollider" that would be by far the biggest and most expensive scientific instrument in history. The proposed machine would be buried in a circular tunnel sixty miles in diameter. It has been dubbed the "Desertron" because it could be built only in the flat deserts of the American Southwest. It will accelerate twin beams of protons to 20 trillion electron-volts and smash them head-on. In the rubble of these titanic atomic collisions, the physicists hope to find clues that will lead them to the "grand unification" of forces and to an understanding of the origin and structure of the universe.

A bold dream, and perhaps an empty one. The guardians of the public purse may decide against spending several billion dollars on a machine that produces particles that live for so short a time that their very existence must be inferred rather than observed directly. But the physicists are hopeful. They have discovered that matter is a thing of astonishing texture and beauty. It is building block and architect, music and composer. It is the head of the snake and the tail, the B of the Beginning and the e of Time. The physicists hope that by probing matter on the smallest possible scale they will be drawn into closer contemplation of the ultimate mysteries that have intrigued theologians, philosophers, and scientists alike. What is the universe? Where did it come from? Where will it go? And what is this thing called life that dances on the surface of matter like a residing flame?

❦

Marie Curie, the brilliant physicist who discovered radium, processed tons of Czechoslovakian pitchblende ore to obtain a fraction of a gram of the new element. She purified the new element and put it into a glass vial. The element she discovered is one of the heaviest in the periodic table, and on that basis the Renaissance alchemist Paracelsus would have assigned it to a place at the foot of Creation. But Marie Curie's vial of radium glowed with an eerie green light. It glowed with a light that evokes the Orion nebula seen through a telescope, or the light of a distant quasar burning at the edge of time. It glowed with the light of Creation itself, the flash of the primeval fireball, the Big Bang. The light in the vial came from an element destroying itself, matter giving itself over to pure energy.

Paul was knocked from his horse by a divine light. Marie Curie was made sick. Even as she worked to release radium from its matrix, her body was bombarded by an invisible and deadly radiation. In the glass vial that glowed with an eerie light was the snake that bites its tail, the light of the angels compressed in a pinch of dust. In her last years Marie Curie suffered the terrible agony of radiation poisoning, racked by chills and fever and the slow destruction of her body. Of her death, her daughter said that "she had joined those beloved *things* to which she had devoted her life, and joined them forever."

In her back-yard shed, shut off from the light of the sun, Marie Curie had stumbled upon the equivalence of matter and energy. Out of this equivalence, cosmologists would come to understand the radiance of the Big Bang, the creation of the stuff of the galaxies, the driving force of the quasars, and the secret of what makes stars burn. Light and matter found a strange unity in the pitchblende dust embedded beneath Marie Curie's fingernails. And the light killed her.

8

STARDUST

O nly a daredevil makes metaphors. To make a metaphor is to walk a tightrope, to be shot out of a cannon, to do aerial somersaults without a net. The trouble with metaphors is that you never know when they'll let you down. You turn a somersault in mid-air, you reach for the trapeze—and suddenly it isn't there.

Take the butterfly, for instance. Surely the butterfly is a safe bet for a metaphor. *The delicacy of beauty. The fragility of life.* The *Oxford English Dictionary* gives half a page to the use of the word *butterfly* to represent all that is "vain, giddy, inconstant, and frivolous." Even Shakespeare does it: ". . . for men, like butterflies, show not their mealy wings but to summer." And there you go, sailing through the air, the daring young man on the flying metaphor, when . . . .

Along comes the mourning cloak butterfly. Maybe as early as February, surely by March, the mourning cloak unlimbers and takes to wing. This year it was February, on a day of exceptional warmth. I was sailing along on my bicycle, patches of snow still on the ground, when—crack!— I collided with spring's first mourning cloak. It must have just emerged from hibernation. A few minutes later, its wings more thoroughly warmed by the February sun, this little flier might have avoided the collision. But its fuselage was still caked with sleep, its struts and hinges stiff, as it lurched into my path. I felt the brittle scales brush against my cheek, deep brown-black scales as brittle as mica, edged in brilliant yellow, dotted with pale blue. Wings three inches across. Where did this delicate thing come from in mid-February?

The mourning cloak is the most prominent of the New England butterflies that winter over as adults. Other species pass the winter in the larval stage, or sealed up tight in egg cases. The monarch migrates to a warmer climate. But not the mourning cloak. When snows fly, the mourning cloak seeks out a hollow tree or a pile of dead leaves and hibernates like a bear. It does not have the bear's fur or the bear's layers of fat. That this fragile slip of a thing survives the killing cold has always seemed to me something of a miracle. But every year, on the first warm day of spring, maybe even during a February thaw, the mourning cloak flutters forth in full-blown black-and-yellow-winged glory. Before other insects have hatched as grubs, the mourning cloak takes to the air.

And there goes the metaphor. Beauty is fragile? Life is fleeting? Not at all. Beauty, it turns out, is tough, and life is well nigh impossible to extinguish. The mourning cloak proves it.

❧

I have to take the word of my handbooks that the mourning cloak butterfly hibernates through the New England winter. I have never found an adult mourning cloak in hibernation. But I have watched the other stages of this remarkable insect's life. The mourning cloak lays its eggs in rings around the twigs of elms, willows, and poplars. Its caterpillar is a fuzzy black-velvet sort of thing, splotched with red and white dots. It rolls itself up into a pupa of brown isinglass, decorated with red tips. The mature butterfly emerges in late summer and feeds on pollen and nectar until time for hibernation. That the mourning cloak finds pollen and nectar when it awakens in February or March is part of the miracle.

In England, where it is prized as a rare migrant from the continent, the mourning cloak is known as the Camberwell beauty. In Europe, too, the mourning cloak announces spring. I will not forget a lovely passage in Nabokov's autobiography, recalling the first blush of young love in St. Petersburg in the spring of 1916, and a "Camberwell beauty, exactly as old as [his] romance, sunning its bruised wings, their borders now bleached by hibernation, on the back of a bench in Alexandrovski Garden." Winter in St. Petersburg is surely more brutal than winter in New England. Yet the mourning cloak survives there too, sleeping in a hollow tree, with wings as thin as tissue paper bravely folded against the bitter cold, to be somehow miraculously revived by the first warm rays of the spring sun to treat the frolickers in public gardens to a fine glimpse of summer's beauty.

The endurance of the mourning cloak is a measure of the versatility of matter. In the mourning cloak's seed case on the willow twig there is not much more than a plan for the future larva, a twist of DNA in a protein wrapper. When the egg is energized by the season, the plan constructs a caterpillar out of atoms taken from the willow leaf and from the air, atoms of carbon and oxygen and hydrogen mostly, with bits of nitrogen and phosphorus and iron and whatever else is needed, atoms stuck here and there in the parts of the growing worm as needed to give a protein shape, body, structure, or an appropriate balance of electronic charge. Then later, inside the pupa, all these same atoms are shuffled again, still guided by the DNA, to become the body of the adult butterfly. Some of the atoms of the adult's body are passed on in the eggs to become part of the next generation of butterflies. And when the adult dies, the final complement of its atoms falls to the ground to become the substance of some other living thing or of earth or of stone.

It would be wonderful to tag an atom of carbon, say, the way an ornithologist bands birds with some sort of tiny transmitter, so that we could track its journey. An atom of carbon is, as far as we know, a permanent thing (provided it is not the radioactive form of carbon known as carbon 14). Every carbon atom that ever was, still is. The carbon atoms of the Earth's crust were once, before the Earth was born, part of the dusty nebulas of space. A carbon atom on the surface of the Earth makes its way around and around like a pilgrim or a gypsy, now into a rock, now into the sea, now into the air, now into the body of a living creature. Its alliances are eclectic. For a while it may join up with a couple of oxygens and travel the roads as CO_2. Or it may take up with a larger crowd of nitrogens and hydrogens and oxygens in the protein of a mourning cloak butterfly. Or it may stick with its own kind in the regimented ranks of a diamond or a block of graphite. And if we had tagged it, banded it, always it would emit the tiny signal announcing its current habitat.

Think of it! Atoms flowing through creation like the wind. The *One and the Many*, the Greeks called it. Behind the shifting flux of things lies the thing that stays always the same. *Everything moves, everything flows*, said Heraclitus. The body of the mourning cloak is like a river. You can't step in the same river twice. The world we live in is a flame; we burn in it, we are burning all the time. The mourning cloak burns like the tongues of the Paraclete, anointing the seasons. The rocks burn with a slow, steady flame. If we could see the flame dancing on the bush, as Moses saw, if we could see every bush, every tree, burning all the time, every twig tipped with flame, the wind, the river, the constant flow of atoms, we would wonder that anything endures. The real constitution of things is accustomed to hide itself, said Heraclitus. The power of the visible is the invisible, said the poet Marianne Moore. "The house, the stars, the desert," repeated the pilot to the Little Prince, having learned this lesson well, "what gives them their beauty is something that is invisible!" It is an old tattoo ringing in the ears of philosophers and poets, physicists and mystics: *the power of the mourning cloak, the resilience of its beauty, what makes it tough, what makes the flame of its elegance impossible to extinguish, is something that cannot be seen.* The surface of the Earth is aflame, and the flame is just a passing about of the enduring thing that cannot be seen. Hydrogen, carbon, oxygen, iron, these are the coin of the realm, the irreducible, permanent, hidden treasury. These are the elements of the flame.

This morning I watched the sky burn, watched matter flowing in the space between the stars. I was up before dawn and I caught in the east the constellations of summer. The stars Vega, Deneb, and Altair were well up before the sky began to brighten. On an impulse, I got out my telescope and went looking for faint lights. First, I found the Ring Nebula in Lyra. In my instrument the nebula looked like a tiny smoke ring, gray and misty. I hadn't bothered to plug in the electric drive of the telescope, so the motor did not compensate for the turning of the Earth. As the Earth sagged eastward toward the sun, the nebula drifted across the field of the eyepiece and reinforced the image of a smoke ring drifting in space. The Ring Nebula is a *planetary* nebula, one of a class of disk-shaped blurs in the sky that early watchers likened to the disks of planets. But the Ring is far beyond the planets; it is 1500 light-years away. It is a bubble of gas blown off of a dying star. The naked core of the star that shed this envelope of dust is invisible in my instrument, but photographs made with larger observatory telescopes show the parent star nested at the center of the bubble like the larva of a butterfly in its bright cocoon.

This is the way stars die. All of their lives stars shed mass. Even now the sun is blowing off an aura of its own substance into space, an ethereal stream of matter called the solar wind. Stars are born of the gas and dust of space when gravity causes nebulas to collapse into dense spheres, and stars give gas and dust back to space from the first day of their lives. For 5 billion years the sun has exhaled a faint breath as it burns, bathing the Earth in the flux of its exhalations, a wind of atoms and subatomic particles that feeds the Earth's atmosphere and ignites auroras. In all of that time the sun has lost only a tiny fraction of its original mass. But toward the end of a star's life, the giving back of substance to space is often accelerated, particularly if the star is a big one. As the nuclear fuels that sustain a star's burning become depleted, the star enters a period of instability—gravity crushes down the core of the star, thermonuclear fusion puffs it up—and the star blows off its outer layers. The stellar convulsion that blew off the Ring Nebula occurred less than 20,000 years ago in a star 1500 light-years away. In another 50,000 years the bubble of gas we see in space will have expanded to invisibility. The smoke ring of star-stuff from the dying star in Lyra will have been dispersed to the interstellar medium.

❦

From the Ring Nebula in Lyra, I swung my telescope down toward the horizon and went looking for the Dumbbell Nebula in Vulpecula. The books say that the Dumbbell is brighter than the Ring, but I always have a harder time finding it. This morning the task was made more difficult by the approaching light of dawn. The nebula takes its name from its shape, "two hazy masses in contact," but to my eye this morning it was no more than a faint oval of pale light. I can't see them with my fourteen-inch telescope, but the books say that there are two stars at the center of the Dumbbell—a binary star system—and as far as I know no one has determined which of the two central stars is the one that gave up its mass to create the nebula. One of those stars suffered a terminal convulsion and blew away its skin. The matter in the nebula is only a small fraction of the mass of the original star. The star shed only its outer layers, the way a cicada sheds its chitin shell.

As the sky near the horizon began to lighten and the Dumbbell faded, I slipped my telescope back to the Ring in Lyra, but almost immediately that nebula too was lost in the glare of the sun. The total mass of the material in the Ring Nebula has been calculated to be about 10 percent of the mass of the sun. The composition of the Ring can be determined from the spectrum of its light. For every

17,000,000 atoms of hydrogen, there are 1,000,000 atoms of helium, 10,000 atoms of oxygen, 5000 atoms of nitrogen, 1500 atoms of neon, 900 of sulfur, 130 of argon, 34 of chlorine, 4 of fluorine. There's carbon too. *And a revelation.*

The relative abundances of elements heavier than hydrogen are *significantly greater* in planetary nebulas than in the surrounding interstellar space; for example, there can be as much as ten times more nitrogen *relative to hydrogen* in a planetary nebula than in the clouds of galactic matter out of which the nebula's parent star was born. And astronomers think they know why. As stars burn, they convert hydrogen to helium, then helium to carbon, and to oxygen, and to nitrogen, and finally to iron. The process is called thermonuclear fusion, and it is the same process that takes place in the explosion of a hydrogen bomb. In the fusion of atomic nuclei at the cores of the stars, a tiny fraction of the mass of the fused nuclei is converted into pure energy, and it is this energy that makes the stars shine. As they burn, the stars build heavy elements from light ones. And the heavy stuff they build they give back to space, slowly in the stellar winds, more rapidly in the expanding bubbles of the planetary nebulas, sometimes catastrophically if the star convulses and blows itself apart as a supernova.

Stars build as they burn. In the workshops that are the cores of stars, gravity squeezes heavy elements into being. Here is work that would be envied by Cellini, more ravishing than Fabergé creations for a czar. Oxygen, sparkling with its six valence electrons, promiscuous in its rage for union, burning, rusting, rotting, building; no element on Earth is more common. Carbon, the wizard; now the black rider, now the diamond throne, the backbone of the butterfly, nylon, gasoline, shoe polish, dynamite, and DDT. Iron, industrious, core of the Earth, night-flyer; Eskimos made tools of iron that fell from the sky. Palladium, zirconium, dysprosium, gadolinium, praseodymium, rare travelers made in traces in supernovas and scattered to the Galaxy like bank notes tossed from a king's carriage.

A hundred generations of stars lived and died in the Milky Way Galaxy before the sun was born, a hundred generations of stars turning primal hydrogen into the stuff of future planets. This morning in the Ring and the Dumbbell I saw the universe spreading bank notes. This morning I saw two stars turning pennies into dollars, turning dollars into gilt-edged certificates. A single atom of carbon is a certificate well worth having, and the Ring and the Dumbbell were spewing them out like confetti.

❧

Every atom of the Earth, excepting the hydrogen and some of the helium, was made in the hot core of a star or in the energetic convulsions that accompany the end of a star's life. Every carbon atom in the graphite of the pencil I write with carries a label that says "Made in Taurus" or "Made in Orion." The universe burns. The atoms flow out of stars and across galaxies—now they are dust in the Horsehead Nebula of Orion, now they are the crust of a new planet—like circulating coins or baseball cards, trading hands. The atoms flow through the body of the mourning cloak and pause only briefly, in and out with every breath. Every second the entire surface of the Earth goes into a pupa and rearranges its form. "The sun is new each day," said Heraclitus. Did he mean it literally? "The thunderbolt steers all things," he said. The universe is afire, kindling in measures and going out in measures. Step in the river and burn your foot. Step again in the same river and burn the same foot and it is neither the same river nor the same foot.

Lyly believed the ostrich digested hard iron to preserve its health. The mourning cloak butterfly digests hard iron. In its belly is beauty. Ten billion years ago that beauty flew out of a star. The power of the visible is the invisible. Stars puff up like smoke rings, like dumbbells,

and we don't see what's important. What's important is
what's going on behind the veil of the nebula, behind the
mourning cloak's isinglass curtain. The mourning cloak's
pupa is the Heavenly City, measured by a golden reed. The
first foundation of the city is jasper; the second, sapphire;
the third, agate; the fourth, emerald; the fifth, sardonyx;
the sixth, sardius; the seventh, chrysolite; the eighth,
beryl; the ninth, topaz; the tenth, chrysoprase; the elev-
enth, jacinth; and the twelth, amethyst. The butterfly slips
into the Heavenly City to change its cloak. Its chrysalis is
gold. It is stardust.

9

FAR DOWN
A BILLOWING
PLAIN

One fine fall day I passed a troop of students from Maura Tyrrell's botany class, marching to their lab with fistfuls of wildflowers. I knew what they would do there. They would open Britton and Brown's *New Illustrated Flora*. They would unfold their bouquets on the illuminated stage of the dissecting microscope. They would count stamens and sepals, examine the colors of corollas, slit open ovaries. They would sketch the form of the plant. Were the flowers arranged in spikes, racemes, panicles, umbels, corymbs, or cymes? Were the leaves dissected, lobed, toothed, or entire? Were they opposite, alternate, or palmate? Was the attachment of the leaves to the stalk petiole, sessile, clasping, or perfoliate? Even as the students marched up the hill to the lab, they snatched up still more plants from the verge of the path—chicory, asters, false Solomon's-seal. I followed them jealously.

The naturalist John Muir said that the two greatest moments of his life were the time he camped with Ralph Waldo Emerson at Yosemite and the time he found the rare orchid Calypso blooming alone in a Canadian swamp. When I first came to New England twenty years ago, Calypso's cousin, the pink lady-slipper, seemed almost as rare. I remember the first one I found, by a pond in the deep woods. I had never seen such a plant in the woods: gorgeous, heavy, tropical, like an escapee from a hothouse. I was dazzled. I sat on the ground next to it and wondered how a plant could make my knees tremble.

It turned out that the lady-slipper was not as rare as I had supposed. As I came to know the area better, I learned where to find them. To come upon a lady-slipper in my woods in those days was perhaps, after all, not quite as momentous as camping with Emerson, but it *was* an encounter to be savored. Now, two decades later, lady-slippers are plentiful hereabouts. You can't walk the woods in early June without plants popping up under your feet. In summer, greenness is cheap, said Thoreau. In their new numbers, lady-slippers are a dime a dozen.

Most of our local wildflowers are as valueless as greenness in summer—winter cress, goldenrods, asters, butter-'n-eggs—they come and go in prodigious numbers. But there are wild plants around here that are rare enough to stop me in my tracks. Once I found a solitary wild columbine and I have never seen another. The stately red cardinal flower grows only in one or two places that I know about, certain intermittent creek beds where the plant is well hidden from the eyes of collectors. And last year I found a plant that might have dropped to Earth from another planet: a *white* lady-slipper, snow-pure, alone in a pine wood with ten thousand of its pink cousins.

My Peterson wildflower handbook admits the white lady-slipper and calls it rare and local. Rare and local! That singular white plant, ordained by uniqueness, was the chosen bride, the anointed lamb. Nothing in Britton and Brown could have prepared me for the thrill of the find. And this season I returned to the same place and the white lady-slipper was there again, miraculously resurrected from the duff of the forest floor, alone among ten thousand pinks. Certain New Guinea tribes have in their vocabularies only two basic color words, which translate roughly as "black" and "white." Two words are enough. My lady-slipper is white. The rest of the universe is black. The galaxies and the stars were created to bring forth this single plant. Stars burned to forge its atoms. *Earth cannot escape the sky,* wrote the fourteenth-century mystic Meister Eckhart; *let it flee up or down, the sky flows into it, and makes it fruitful whether it will or no. So God does to man. He who will escape him only runs to his bosom.*

❦

At least 3 million and perhaps as many as 10 million species of life now exist on the surface of the planet Earth. A much greater number of creatures existed in the past, but many have become extinct. This is fruitfulness on a scale somehow unseemly for a small planet near a backwater star. Nothing else we know about the universe matches the Earth's diversity of life. The physicist recognizes a few dozen elementary particles and ninety-two naturally occurring elements. The astronomer catalogues galaxies in a system involving fewer than twenty categories. Stars are classified in an arrangement with fewer than a hundred refinements. And here on the heaving, thrashing, palpitating surface of the Earth there are 10 million separately distinguishable forms of life. There are not enough biologists alive to make sense of it all. The white lady-slipper can be placed rather precisely in the botanist's scheme of things: kingdom Plantae, phylum Angiospermophyta, class Monocotyledoneae, order Orchidales, family Orchidaccae, genus *Cypripedium*, species *acaule*. But even that detailed declension is not good enough, for the same botanical litany applies to the far more common pink form of the plant. What kind of universe is it that so runs riot? What kind of universe will make room for a single white plant when a hundred thousand pink ones are more than enough? It is a huge question, perhaps the roomiest question we will ever ask.

Cosmologists and physicists are asking the same question more and more frequently—and they have reason to. The more we learn about this particular universe we live in, the more impossibly improbable it seems that we are here at all. Consider these coincidences: If the so-called fine-structure constant that governs atomic interactions were even slightly different from its known value, stars would either burn out very rapidly or remain forever cold and dark. The value of that constant is balanced like a penny on its edge in such a way that there is a sizable class of stars (including the sun) that will burn with a steady light for billions of years, and therefore heat and nurture developing life on a nearby planet. If the strong-interaction constant that governs the nuclear force were just 2 percent larger, it is unlikely that protons could form from their constituent quarks, and there would be no elements as we know them. If the constant were a few percent smaller, the nuclei of elements heavier than helium would be unstable, and there could be no such thing as a planet made of rock and iron and covered over with a gauze of carbon-based life. If the gravitational constant differed slightly from its observed value, plus or minus, stars could not shine long enough for life to evolve on nearby planets. If, one second after the Big Bang, the ratio of the density of the universe to its expansion rate had differed from its assumed value by only one part in 10^{15} (that's 1 followed by fifteen zeros), the universe would have either quickly collapsed upon itself or ballooned so rapidly that stars and galaxies could not have condensed from the primal matter.

The bare fact of my white lady-slipper's existence sets severe constraints upon the values of the fundamental constants of physics. That plant could not have come to be—nor you or I—unless the constants governing the laws of nature had had certain extraordinarily precise and apparently gratuitous values. Even the fact that the stars shine selects one universe out of a possibly infinite number of imaginable universes. The coin has been flipped into the air and come down on its edge. The cosmologist is as baffled by all of this as you or I. He has raised the improbability of this universe to a principle, the *anthropic principle.* The anthropic principle states that since the only universe we could possibly observe is one that has the qualities that allow for the appearance and evolution of life, then the universe we live in must *necessarily* have those qualities. The coincidences in the physical constants may seem to have been miraculously contrived for our benefit, but *since we are here to observe them* they could not have been otherwise.

Blake was right to see the world in a grain of sand and heaven in a wild flower. The silicon and oxygen in the grain of sand and the carbon in the flower could not have come into being unless the forces that hold the universe together had exactly the values they do. Adjust the strength of the electromagnetic force or the nuclear force but slightly, and you knock out of kilter the resonance in the carbon nucleus that allows three helium nuclei to come together in the cores of stars to form that element. Stop the synthesis of the elements at helium, and never in a billion years of burning would a galaxy of stars produce enough silicon or oxygen to make a single grain of sand. No, the coin did not come down on its edge. The situation is more improbable than that. The coin was flipped into the air 10^{15} times, and it came down on its edge but once. If all the grains of sand on all the beaches of the Earth were possible universes—that is, universes consistent with the laws of physics as we know them—and only one of those grains of sand were a universe that allowed for the exist-ence of intelligent life, then that one grain of sand is the universe we inhabit.

There are those who will applaud this information and say, "Ah, you see, the physicist has proved that God exists." And I say, "no such thing"; the physicist has proved nothing. He only observes that this wildly improb-able universe exists, and it is the only universe we could possibly observe. If that is a mystery that holds us in thrall, then so be it. If you wish, I will give praise to this improbability in the language of my forefathers and with cymbals and trumpet make a joyful noise. I will offer as an unblemished sacrifice the one white flower in the piney woods, but I will not for a moment assume that the lan-guage of my forefathers encompasses this new knowledge of the Infinite. What the physicist has learned is no less frightening, or less wonderful, than the mysteries that drove anchorites to desert rocks and Buddha to the Bo tree.

What the physicist has learned enriches and deepens those venerable mysteries; it neither proves them nor negates them.

❦

It does indeed look at first glance as if the universe were designed expressly for our benefit. But there is another way of looking at it. The quantum physicist, when he turns his discipline to the physics of the Big Bang, tells us that the equations allow for multiple universes to pop into existence like bubbles: an infinite number of bubble universes, springing into existence when the cork is popped in some kind of hyperspace and hypertime, and in every universe physical constants and conditions assume random values. The universe that we know, this universe of sand and white flowers, is only one of the class of universes that allows for the formation of our kind of stars and our kind of matter. The coin that is tossed 10^{15} times comes down in every possible configuration, and our night is just one of 10^{15} nights, some crammed with stars, some forever dark, and all but one of which we shall never know. Listen to Alan MacRobert writing in *Sky & Telescope*:

> After 50 years, the Big Bang universe of galaxies seems a little constricted, naggingly inadequate. We seem to be approaching the next step outward. We peer today as through a barely open door at a prospect of universes without end, great and small, familiar and incomprehensible, in numberless profusion. Valid physical questions face us for which our physics is utterly inadequate. This can only be a sign that we stand at a great frontier of science, one that will form a cutting edge of inquiry for generations to come, with results we cannot guess.

What am I to make of this? Universes as numerous as the bubbles that fizz from champagne! And I am barely getting used to this one. A single starry night is enough to make my head spin. One white orchid takes my breath away. Who am I to say if God exists or if this universe of galaxies is only a bubble of cosmic foam? In the night even my lady-slipper is black. I am a pilgrim of darkness. Like the poet Roethke, "I look down the far light / And I behold the dark side of a tree / Far down a billowing plain, / And when I look again, / It's lost upon the night."

❦

Today the newspaper said that the space shuttle *Columbia* would pass over our area at 4:47 P.M., just after sunset. At 4:45 I went into the yard to see what I could see. Exactly on schedule, the shuttle appeared in the northwest, moving with the precision of a surgeon's scalpel toward the east. It was as bright as Venus, like a planet set loose from its ancient track. As I stood watching the silent, distant craft make its clean incision on the sky, suddenly the night was full of birds. A flock of Canada geese took up from the pond to the north of the house and churned south, not cutting the night so much as thrashing it, beating the night into the ground with the great thump-thump of their wings, and hooting the stars from the sky with their throaty honks. Forty or fifty geese, in a ragged vee, not a hundred feet above my head, beating, thrashing, flowing south with a terrible racket, a noise of ram's horns and shouts that would have toppled the ramparts of Jericho. I never thought to look to see where the shuttle went from sight.

The surface of the Earth flows with life. Everywhere I look the place is rank with it. I open my eyes and the night is full of birds. I unstop my ears and I hear the beating of wings. What sort of profligacy is this? On my bookshelf is Margulis and Schwartz's guide to the phyla of life on Earth, a 270-page classification boldly embracing all that lives. You and I and the wild geese, and every other creature with a brain and a spinal cord, 45,000 species of us, have four pages—236 through 239; the lady-slipper and all other flowering plants, 230,000 species, have three pages—268 through 270. I am astonished at how few of the species in the book I even recognize, whole armies of creatures too small to be seen without a microscope or residing in habitats where I will never go. I turn over a rock and a thousand animals scuttle out of sight. I scratch my eyelid and disturb a nation of mites. With every step I topple microscopic forests. I open the book at random, to page 218: the Tardigrada, the slow-stepping, mite-sized "water bears." And here is a photograph of a tardigrade made with a scanning-electron microscope. One can see why Huxley called them "bears"; they look like bears and move like bears, although reduced in size 10,000 times. The tardigrade can survive temperatures as high as 150 degrees Centigrade and as low as 3 degrees above absolute zero. They live in the arctic and in the tropics, in fern forests, and in hot springs. They can turn themselves into dry, barrel-shaped forms—called *tuns* because they resemble wine casks—and, in this state, survive for a hundred years. They can endure radiation levels a thousand times stronger than those that will kill humans. And this whole tribe of creatures goes unseen, climbing across lichens on microscopic bear paws or stacked in dark cellars at the roots of plants like casks of aging wine. Should I be astonished, then, if it turns out to be true that in the ensemble of all universes this night of the honking, thrashing geese is only one of many? Is it really so unbelievable that nature would create universes in limitless numbers—universes doomed to exist unseen because they lack sentient creatures—in order to ensure that one or two or a hundred would randomly emerge with exactly the right fluctuations of physical constants that permit a white flower in a piney woods? If 10 million species of life exist on Earth when one would do, if a hundred billion galaxies exist when one is enough to ensure that a thousand planets will be clogged with life, then why not a trillion billion universes? Why not, indeed! The thumping and hooting of the geese, hours later, still disturbed the night.

10

HIDDEN MATTER

O n the night of the peak of August's Perseid meteor shower, my son and I slept under the open
sky. It was a night of exceptional clarity, far
from the lights and haze of the city. Meteors
flashed against a background of stars so numerous that the
heavens seemed more light than darkness.

Above our heads the Milky Way arched from Cassiopeia in the north to Sagittarius in the south. It was a dark-
shoaled river of light, a luminous drapery, diamond dust
on black velvet. As the hours passed and the stars tilted
toward the west, we could almost feel ourselves whirled in
the bright vortex of the Galaxy.

For the ancients of all cultures, the Milky Way—the
Via Galactica—was a bridge, a road, or a river. The astronomer Robert Burnham has suggested that these images are
closely related to the idea of human life as a journey or
voyage between two worlds—*life is a bridge,* says the Zen
master, *build no house upon it*—and to the endless journey
of the universe itself toward an unknown destination. It is
a powerful, evocative symbolism that goes a long way toward making us feel at home among the stars.

The modern view of the Milky Way is no less evocative. Our sun is one star in a disk-shaped swarm of several
hundred billion stars. The disk is 100,000 light-years in diameter. The stars in the disk are clustered in spiral arms
that turn about the center like a pinwheel; a star at the
sun's distance from the center (30,000 light-years) makes
the circular journey about the axis every 250 million years.
The space between the stars is strewn with dust and gas,
and it is out of this rich reservoir of material that the stars
are born. Above and below the galactic disk are scattered
several dozen spherical swarms of millions of stars called
globular clusters. The globular clusters buzz about the Galaxy like bees about the hive.

I have often constructed a model of the Milky Way
Galaxy on a classroom floor by pouring a box of salt into a
pinwheel pattern. The demonstration is impressive, but
the scale is wrong. If a grain of salt were to accurately represent a typical star, then the separate grains should be

thousands of feet apart; a numerically and dimensionally precise model of the Galaxy would require 10,000 boxes of salt scattered in a flat circle larger than the cross-section of the Earth. On the scale of the Milky Way Galaxy, our solar system is a mote of dust riding a whirlwind, a grain of salt in a salty sea. All of the stars we saw that star-powdered night my son and I slept under the open sky, the thousands of stars that can be seen on the best of nights, are but our nearest stellar neighbors in one arm of the circular tide.

The brilliance of the summer Milky Way in the direction of the constellation Sagittarius hints at the great density of stars that cluster near the center of the spiral. Dust and gas in the central plane of the galactic disk obscures our view of the nucleus, but radio and X-ray telescopes have penetrated the obscuring matter and revealed a powerful source of energy, a monstrous pulsing heart for the Galaxy, a core of violence that recapitulates the violence of the Creation itself. The nucleus of the Milky Way Galaxy is apparently the site for cosmic convulsions on the grand scale, perhaps a place where countless suns are swallowed up by a massive gravitational black hole.

❧

When Galileo turned his telescope upon the Milky Way in the winter of 1610, he was astonished to see that band of pale light resolved into stars, stars apparently uncountable in number, individually beyond the limit of unaided vision. It is the collected light of these myriad stars in the plane of the galactic disk that we see as a luminous river encircling the celestial sphere. The concentration of obscuring dust and gas in the central part of the disk, especially in the direction of Sagittarius, breaks the stream of the Milky Way with islands of darkness. So bright was the Milky Way in earlier, darker times that people who lived in southern latitudes (the latitudes of Sagittarius) made dark constellations of the rifts that interrupt the river of light. Australian aborigines saw the enormous figure of an emu in the dark regions of the central Milky Way, gulfs of darkness that we somewhat less poetically call the Coal Sack. The Quechua-speaking Indians of Peru recognized several dark constellations in the gaps of the Milky Way, including a bird, a fox, a baby llama, a toad, and a serpent. Spanish conquistadors dismissed the Inca descriptions of dark figures; the starless voids did not correspond to the European idea of what a constellation should be. Where Europeans saw broken light, the Indians saw bounded darkness. Where Europeans saw the road, the Indians saw the beasts upon it.

And now the dark beasts of the Incas have come back to haunt us. In recent years astronomers have discovered that the greater part of the contents of the Milky Way has previously eluded us, hiding in the darkness between luminous stars like the Inca llama or toad.

Radio studies of the emissions of atomic hydrogen show that the disk of the Galaxy is about twice as extensive as had been previously thought. Clouds of hydrogen extend well beyond the region of active star formation, rimming the star-bright Galaxy with tentacles of gas, four starfish arms that reach into the void of intergalactic space, doubling the size of the spiral with streamers of matter too tenuous to give birth to stars.

Another unexpected extension of the Galaxy followed from the simple business of counting stars. Stars are now counted on photographic plates with computer-controlled scanning light beams. These new techniques have enabled astronomers to catalogue stars a hundred times fainter than in previous reliable determinations. They have discovered a huge number of stars of low intrinsic brightness distributed above and below the galactic disk in space thought to be devoid of matter. The Milky Way spiral is apparently enclosed in an egg-shaped aura of faint stars.

But the most startling development followed from studies of the rotational dynamics of the Milky Way Galaxy. Observations of the motions of objects in the outer reaches of the Galaxy—bright stars, clusters of stars, and clouds of molecular gas—have enabled astronomers to refine their calculations of the Galaxy's total mass. The orbiting motion of these objects is controlled by gravity. They are held in orbits about the galactic center by the gravitational pull of all of the mass that lies closer to the axis of the galaxy than the objects themselves. By observing the velocities of these distant objects, astronomers can calculate the quantity of mass that is holding them in orbit. The result of these calculations was a surprise. It turns out that the Milky Way Galaxy contains at least thirty times more mass than had been previously estimated, nonluminous mass that manifests itself only through the ineluctable pull of its gravity.

How could so much matter, 97 percent of the massy substance of the Galaxy, have gone undetected until now, hiding in darkness the way the constellations of the Incas eluded the light-struck Spaniards? The answer is that the newly discovered mass must consist of some exotic forms of nonradiant matter. Not stars, which we should certainly see ablaze in the heavens. Nor interstellar gas and dust, which we would observe by their own radiation or by the absorption of light from distant sources.

What, then, is this hidden component of the universe, this dark matter that constitutes the greater part of all that exists? I once recounted these developments to a friend. "Ninety-seven percent of the stuff in the universe," I said, "is stuff about which we know absolutely nothing." "It is probably the best stuff, too," my friend replied. The truth is that astronomers do not as yet have any idea what this "stuff" is that holds the stars in their galactic orbits.

Perhaps the dark component of the Milky Way Galaxy consists of Jupiter-sized objects, spheres of hydrogen and helium too small to have "turned on" as stars. Perhaps it consists of some as-yet-unknown type of celestial body, larger than grains of dust but smaller than the smallest stars. Or perhaps the dark mass consists of black holes, stars that have collapsed into themselves and become so dense that light cannot escape the pull of their gravity.

These are the more conventional candidates for the invisible matter that pervades the Galaxy. Other forms of "dark stuff" have been suggested by the physicists who investigate the realm of the subatomic: hoards of neutrinos, each endowed with an imperceptible whiff of mass; or a gas of yet-to-be-discovered "gravitinos" or "photinos" or "axions," particles a trillion times lighter than electrons, hypothetical entities so bizarre and remote from ordinary experience that no one could have thought of them who did not wander like a pilgrim among the modern kingdoms of Prester John, the worlds of the infinitely large and the infinitely small. Like the atoms of the eighteenth-century philosopher Roger Boscovich—particles with zero dimension and an infinite field of influence—these outlandish beasts on the road of the Milky Way may bind the universe together with a thread so ineffable that it can be comprehended only by letting our imaginations embrace the entire fabric of space and time. *Canst thou bind the sweet influence of the Pleiades,* the Lord answered Job from the whirlwind, *or loose the bands of Orion?*

❦

What is this marriage of the infinitely large and the infinitely small that we toss like a net over the stars? Is it an intuition of a Grand Design or only a projection of our dreams? Man has always seen his dreams in the Milky Way. To the Egyptians the Milky Way was wheat spread by Isis; to the Eskimos, a fall of snow. Bushmen saw campfire ashes; the Arabs, a river.

Shortly before his death at the age of eighty-seven, Harlow Shapley visited my college. Shapley was the man who gave us our modern version of the Milky Way. It was he who proposed the pancake spiral of a hundred billion stars, a hundred thousand light-years in diameter, streaked with bands of dark or glowing gas. The headline in the *Boston Sunday Advertiser* on May 29, 1921, had read: UNIVERSE THOUSAND TIMES BIGGER, HARVARD ASTRONOMER DISCOVERS. On his visit, the frail octogenarian told us the story of his life and what it meant. Eyes still bright, white hair askew, he raced across the Milky Way and left us mere citizens of the Earth agog.

Shapley dreamed a grand dream. His dream encompassed all of the visible stars. His dream enlarged the universe a thousand times. And now we discover that the visible stars are only a small fraction of what exists. Most of the mass of the Milky Way Galaxy is invisible to us, lying in the darkness at the edge of dreams, holding, binding, shaping, directing—Leviathan hidden in a sea of light, or ponderous Behemoth with bones like pipes of brass and gristle like plates of iron. If the newly discovered nonluminous matter of the Milky Way is typical of other galaxies (and there is reason to believe that it is), we will have to drastically increase our estimate of the mass of the universe. According to the theory of general relativity, it is the average mass density of the universe that will determine whether the universe will either expand forever into a cold void or fall back upon itself to recreate the blazing fireball of its creation. The hidden matter of the Milky Way and other galaxies may decide the ultimate fate of the universe.

❧

All of this my son and I knew and felt that August night when we watched the annual display of "shooting stars," adrift on our backs in the stream of the Milky Way. We rode a spinning Earth about a sun that spun in a whirlwind of a trillion stars, sealed in an envelope of mysterious dark matter, wheels within wheels within wheels, sharing the journey of the universe itself toward an uncertain destination.

11

THE MONSTER
IN THE POOL

O dysseus had Circe's warning. On the one hand he would encounter the monster Scylla, the beast with six heads and six gaping mouths, each mouth set thick and close with triple rows of teeth. On the other, he would pass the whirlpool Charybdis, who three times each day sucked the waters down and spewed them up again. Charybdis seethed like a cauldron on a fire, and the spray she flung rained down on every side. When she gulped the salty sea the whole interior of her vortex was revealed, and the dark sands at the bottom came into view. Odysseus made his choice. He steered near to Scylla, and when the six-headed monster snatched the six best men from the deck of his ship he counted himself fortunate to have escaped with the rest past terrible Charybdis.

Homer's Scylla anticipated the Great Kraken, which Bishop Pontoppidan, the eighteenth-century prelate of Bergen, described in his *Natural History of Norway.* It is an account less given to the embellishments of myth than Homer's story, but still an ancient magic works upon the bishop's imagination to cast the Kraken in ominous airs. The body of the Kraken is so broad and great, maintains Pontoppidan, that fishermen find themselves in unnatural shallows when the monster floats just beneath the surface of the sea. The creature rises and sinks, its body is full of thick slime like a morass, and its arms are as long as the masts of ships. If the Kraken's arms were to lay hold of the largest man-of-war, they would pull it to the bottom. "After the monster has been on the surface of the water a short time," writes the bishop, "it begins slowly to sink again, and then the danger is as great as before, because the motion of this sinking causes such a swell in the sea, and such an eddy or whirlpool that it draws down everything with it."

Herman Melville, writing in less credulous times, was inclined to believe that the inspiration for the bishop's monster was the giant squid. In *Moby Dick*, the whalers on Ahab's *Pequod* encounter a giant squid and briefly mistake it for the white whale. Ishmael's account is not easily forgotten:

> *Almost forgetting for the moment all thoughts of Moby Dick, we now gazed at the most wondrous phenomenon which the secret seas have hitherto revealed to mankind. A vast pulpy mass, furlongs in length and breadth, of a glancing cream-color, lay floating on the water, innumerable long arms radiating from its center, and curling and twisting like a nest of anacondas, as if blindly to catch at any hapless object within reach. No perceptible face or front did it have; no conceivable token of either sensation or instinct; but undulated there on the billows, an unearthly, formless, chance-like apparition of life.*

The sailors shuddered. All but Ishmael would one day be whirled into the depths to join the monster.

Since ancient times, the monster and the whirlpool have been stock items in myth and story. And, as might be expected, those images have made their way into the sky. There is a whirlpool at the foot of Orion, near the bright star Rigel, called "death" by Hermes Trismegistus and "the way to Hades" by the Maori of New Zealand. From this place among the constellations flow the waters of the river Eridanus, the watery grave of Phaethon, a starry stream that cascades away beyond the southern horizon to the underworld. Near the stream is the constellation Cetus, the whale, the sky's Moby Dick. There is some confusion in the myths of the sky between the river Eridanus and the starry stream of the Milky Way; both sometimes served as the passage between the world of the living and the world of the dead, both harbored monsters, and both were said to have received Phaethon in his fall. In the instance of the Milky Way, science and myth have a marvelous confluence, for we have discovered that the luminous river in the sky is indeed a whirlpool, and in the depths of that maelstrom, even as the ancients guessed, lurks a frightful monster.

❦

Astronomers have known for a long time that the center of the Milky Way Galaxy is the seat of ominous events. The nucleus of the Galaxy is located in the direction of the constellation Sagittarius, and Sagittarius is ablaze with beacons of radiant energy: radio energy, X-ray energy, and ultraviolet energy. Half a century ago a young engineer named Karl Jansky constructed a rotating radio antenna to study the arrival of shortwave static during thunderstorms. Some of the static detected by his antenna came from nearby storms, some from distant storms, and some from a constant faint source that moved completely around the horizon once each day. After several years of persistent observation, Jansky concluded that the faint source of radio static was located in the direction of Sagittarius—beyond any terrestrial storm. The source was fixed in the space of the stars and far removed from the turning Earth.

Jansky's antenna was the first radio telescope. His observation of the source of radio static in Sagittarius was the beginning of radio astronomy. Later, more highly developed radio telescopes were used to map the spiral arms of the Milky Way Galaxy, arms made of stars, dust, and gas, humming with the microwave music of atomic hydrogen. Those same radio telescopes revealed the core of the Galaxy to be a powerful source of radio emission. The most intense focus of this emission, at the very center of the spiral, is called Sagittarius A. The source of the emission is hidden from ordinary optical telescopes by obscuring dust and gas that populate the spiral arms of the Milky Way, but radio waves from the galactic center penetrate the haze as readily as local radio broadcasts pass through the walls of my house. At frequencies of millions of cycles per second, the monster at the heart of the whirlpool announces its presence.

The most detailed radio map of the center of the Milky Way Galaxy has been made with the Very Large Array (VLA) radio telescope, a set of twenty-seven dish-shaped antennas spread out in a Y pattern over twenty miles of desert near Socorro, New Mexico. The radio image shows three S-shaped lobes of hot, electrically charged gas spiraling around the central core of the Galaxy. Astronomers are now convinced that the source of these intense emissions is matter streaming into a massive central black hole. A black hole is an object of such exceptional density that not even light can escape the pull of its gravity. The black hole at the center of the Milky Way Galaxy has a mass equivalent to that of a hundred million suns, squeezed into a volume of space smaller than that of the orbit of the Earth. Apparently, the black hole at the center of the Galaxy continues to grow, drawing stars, dust, and gas into the inexorable sphere of its influence. Not even Odysseus could have dreamed of a monster more bizarre than the object that resides at the center of the Milky Way spiral, a kernel of matter dense and dark beyond imagining, feeding upon stars.

❦

One day in the autumn of 1983 I passed a group of young fishermen on the plank bridge where a local brook widens to form a broad, dark pool. They were casting their lines into the far corners of the pond with the kind of unstudied confidence possessed only by twelve-year-old boys. I stopped to admire their deft skill and moan for the fate of the fish. A few bony shiners lay glassy-eyed and open-mouthed on the planks of the bridge. "How's fishing?" I asked. There was a seriousness in the quick glances they allowed me; they had business to attend to and I was an interruption. "Somewhere out there is a sixteen-inch bass," said one of the boys, dipping his rod toward the far side of the pool. I understood. One after the other, the boys tossed their lures into whatever shadow might hide the old fish. I strained my eyes but saw nothing but black water.

Each afternoon as I walked home from work the boys were at the bridge. Each afternoon I asked for and received the same reply: "He's still there." I could admire the patience and purpose of the boys, but a deeper sort of admiration grew for the scaly creature I had never seen. Like a Loch Ness monster or the Kraken in the fjord, the old bass lent the pond—the world, even—a new enchantment.

Every dark pool should have its monster. The dark pool of the night sky once harbored its share of monsters: Cetus, Draco, Scorpio, and Hydra still lurk among the constellations. For generations we have been busy rooting out the last of those darkling creatures that used to lie everywhere just below the surface of credulity, like the Kraken in the false shallows of Norwegian seas. No modern bishop could write with such a mixture of scientific objectivity and moral foreboding about a flotilla of squids as did the eighteenth-century bishop of Bergen. Some months, says Pontoppidan, the Kraken eats, and other months it voids its excrement. During the months of evacuation the surface of the sea is muddied with the creature's refuse, and the turgid stuff yields a taste or smell that attracts fish. Fishermen come in pursuit of the fish, in their open boats above the Kraken, which opens its arms and seizes and swallows its welcome guests.

One by one, the dark pools of the Earth and sky have given up their monsters to the dragonslayer of science. But now, in a curious returning, science has given us back a monster in the sky stranger than any we might have dreamed. The Milky Way, the whirlpool of the night, has a monster at its core. Stars and planets, gas and dust are sucked to extinction in that terrible vortex, down into an underworld of irrefutable gravity. And like the old bass in its lair, the monster at the heart of the Milky Way deepens and darkens night and casts upon the constellations a spell of magic.

❧

If we want to understand the Milky Way, it is usually best to look to other galaxies. The Milky Way is the one galaxy we cannot see in its entirety, for we are inside it. But if we look up and down out of the flat spiral of our own galaxy we can see countless others, and there is no reason to believe that the other galaxies are unlike our own. We know that the center of the Milky Way Galaxy is the site of mysterious and extraordinary events producing prodigious quantities of energy. The source of that energy, according to astronomers, is a massive black hole that lurks at the center of the starry whirlpool like the Kraken in its vortex. The best evidence for a black hole at the center of our galaxy comes from beyond the Milky Way, from the spiral galaxy known as NGC 4151 in the constellation Canes Venatici.

NGC 4151 is one of the brightest of the Seyfert galaxies, a class of rare, highly energetic galaxies identified by Carl Seyfert in 1943. Astronomers believe that the Seyferts are related to the distant and mysteriously luminous quasars, or at least that the bright central regions of these strange galaxies are powered by the same mechanism that illuminates the quasars. And it now seems likely that the engines at the centers of the Seyferts, and by implication at the centers of quasars, are black holes with masses hundreds of millions of times greater than the mass of the sun. Such an object at the center of a galaxy would draw in and swallow up the stars, dust, and gas in its neighborhood. Millions upon millions of suns, along with whatever planets might orbit those suns, would be whirled to oblivion in that dread sink, falling like ships and sailors into the mouth of Charybdis. As the doomed solar systems and their attendant gassy clouds were drawn into the vortex, they would be spun into a flat "accretion disk," a swirling ring of stars, dust, and gas, piling up, heated, turbulent, blazing with radiant energy before disappearing forever into the fixed oblivion of the black hole.

A team of European astronomers has obtained solid evidence for a black hole at the center of NGC 4151. Using an ultraviolet telescope on an Earth-orbiting satellite, they found in the light of the galaxy's active nucleus the radiation of carbon and magnesium. The characteristic wavelengths of the light were blurred, presumably by the motion of the matter as it orbited the central black hole. From the degree of blurring, the group was able to calculate the orbiting velocities of the clouds of carbon and magnesium.

During a series of observations in 1979, the nucleus of NGC 4151 flared up sharply. Thirteen days later the carbon light flared, indicating that the clouds of luminous carbon were thirteen light-days from the center of rotation. Thirty days later the magnesium flared, giving a distance of thirty light-days from the center for the clouds of magnesium. Knowing the distance and velocities of the orbiting clouds of matter, the astronomers found it an easy task to calculate the size and mass of the object in the middle. At the center of NGC 4151 there is an object not much larger than a single star with a mass equivalent to that of one hundred million suns. There is little doubt that an object of this size and mass must be a black hole. One hundred million suns, perhaps one hundred million planet systems similar to our own, have been whirled to extinction down that cosmic drain. *Heaven keep you from that spot*, warned Circe, *for not even the Earthshaker could save you from disaster.*

❦

J.R.R. Tolkien, the Master of Middle Earth, suggested that the world of magic and monsters is only another view of adjectives. The mind that thought of "light," "heavy," "gray," "yellow," "still," and "swift" also conceived of the magic that let heavy things fly and turned lead into gold and stones into running water. By contrast, the world of the twentieth century is by and large a world of nouns and verbs, a world of objects moving in space and time, a world devoid of magic. What we see is what we get.

Except . . . for that time each afternoon at the pond where the boys fished. From adjectives, I let my mind invent the old bass in his lair. He was as light as a cloud in that watery element or lay on the bottom like a smooth stone. He hid in the gray shadows of the pickerelweed or flashed yellow like refracted sunlight. He was as still as the reflection of the moon in the water or as swift as the whirligig on the glazed surface of the pond. Adjectives are the stuff of the imagination.

Thirty thousand light-years from the sun, at the center of the Milky Way spiral, a gargantuan black hole goes about its business. Heavy, black, and still, the gravitational singularity sits at the heart of darkness, feeding upon the light, the yellow, and the swift. Not even the credulous Pontoppidan could have imagined such a creature, a hundred million stars squeezed into a sphere the size of the Earth's orbit. No, it is the *black hole* that is the size of the Earth's orbit, the volume of space forever cloaked by gravity; the stars themselves, the hundred million stars, are squeezed down *inside* the black hole to the size of the Earth, to the size of a baseball, to the size of a pinhead, to a density ultimately infinite and compelling, creating such an eddy or whirlpool that it draws everything down with it.

The black hole, the white whale, the monster in the pool: The image won't go away. In the horrific conclusion of Melville's moral fable, Moby Dick pulls the universe down after him into the inky depths.

For an instant, the tranced boat's crew stood still; then turned. "The ship? Great God, where is the ship!" Soon they through dim, bewildering mediums saw her sidelong fading phantom, as in the gaseous Fata Morgana; only the uppermost masts out of the water; while fixed by infatuation, or fidelity, or fate, to their once lofty perches, the pagan harpooneers still maintained their sinking look-outs on the sea. And now, concentric circles seized the lone boat itself, and all its crew, and each floating oar, and every lancepole, and spinning, animate and inanimate, all round and round in one vortex, carried the smallest chip of the Pequod out of sight.

❦

On warm autumn nights, when the Milky Way drapes its pale circumference of light from Cassiopeia to Sagittarius, I remember the monster at its core. No Scylla or Charybdis could be more magical than this strange creation of the physicists, revealed by the instruments of the radio astronomers: Sagittarius A, the black hole that feeds on suns, lying still in the pool of night.

In late October, at the pool where the boys fished, at last I saw the old bass. Just as I stepped onto the bridge, he darted from the place in the water lilies where he had been taking the sun. The weather was cold; the boys with their fishing rods were gone. The light and the heavy, the gray and the yellow, the still and the swift, had survived for another season.

12

NIGHT BROUGHT
TO NUMBERS

The mass of men lead lives of quiet desperation, said Henry David Thoreau, and if his book continues to attract us it is because we are desperate. In desperation, I turn to night as Thoreau turned to his pond. I measure those starry spaces with the same care of rods and chains that the naturalist of Concord used to measure Walden. Thoreau plumbed the depths of Walden Pond and marked them on his map. He surveyed the fish that lived in the waters of the pond, he catalogued its weeds, and during winter he recorded the thicknesses of the ice. It was a part of his balance book, an accounting of his riches, a reckoning of a fortune that was there for the taking. These, said Thoreau—the measures, the depths, the thicknesses—are a man's true economy.

The night is my pond. I count stars. I can look out the window and tell you within a deuce how many stars I will see within the Great Square of Pegasus when that constellation rises an hour from now. Tonight it will be only four. Four, or none, or the twenty or so stars that can be seen within the Square of Pegasus on the best moonless nights—the frame is what's important. The Square of Pegasus looks out on space enough, a space fenced with light-years and tracked with the trails of comets. Thoreau did not feel crowded, though his house was within earshot of Concord. There was pasture enough, he said, for his imagination. The low shrub-oak plateau on the opposite shore of Walden Pond stretched away toward the prairies of the West and the steppes of Tartary. Where he lived, he said, was as far off from Concord as the regions viewed nightly by astronomers. He could imagine rare and delectable places beyond Cassiopeia's chair (I use his words), but he discovered that his house on Walden Pond had as its site a place as withdrawn from the life of the town as the stars of Cassiopeia are removed from the sun, "dwindled and twinkling with as fine a ray to my nearest neighbor, and to be seen only in moonless nights by him."

Tonight I will have only four stars in the Square of Pegasus. And 400 others scattered randomly across the night sky. I husband them. A meteor streaks the sky, and

I record it. I weigh out nebulas. I dam up the Milky Way and use it to grind my grain. I put up summer stars like vegetables in jars for my delectation in winter. I have winter stars folded in boxes in the attic for cloudy summer nights. I go to the stars with my ledger open. I count credits and debits. This is an economy, for all it's worth.

❧

Thoreau was much interested in economy. He catalogued and measured with an accountant's care. We know exactly what his house cost: Eight dollars and three-and-a-half cents for boards, four for shingles, and one twenty-five for lath. He spent four dollars for a thousand old bricks, and a penny for chalk. Altogether, his expenses came to twenty-eight dollars and twelve-and-a-half cents, and part of that he spent grudgingly. He needed little but what he could see from his door. He would rather sit by himself on a pumpkin, he said, than be crowded on a velvet cushion.

I read somewhere that a member of the Kung tribe of the Kalahari Desert owns material possessions weighing an average of twenty-five pounds. I figure that my own worldly goods, excluding dwelling, would tip the scales at something more than four tons. That's a heavy burden to lug through life. The acquisition and maintenance of those four tons are the source of my desperation. Thoreau said that most of the luxuries of life, and many of the so-called comforts, are not only dispensable, but also positive hindrances. I have no velvet cushions, but there are hindrances enough around my house, objects that take more than they give, objects that must be carried about whether I wish it or not.

If I had to reduce my load to twenty-five pounds, what would I keep? My first thought would be my bicycle—sleek, black, and mechanically perfect—a prized possession, and surely an acceptable accouterment for middle age. But the trim alloy bicycle weighs twenty-seven pounds, and there are no accessories I can strip off. Besides, winter is coming on, and the bike won't take me through the snowy woods. So how about my stereo and a few favorite records, perhaps Chopin's *Nocturnes*, or Mahler's *First Symphony*? But these, I find, weigh thirty-eight pounds, and that's with the headphones rather than the speakers, and the need for electricity would be still another link to the world of things. Well, then, shall I keep my writing table? That handsome board has given me many moments of pleasure and takes little in return. But my joy in the table is tied to its place in the big bay window with winter sunlight streaming in through green plants and coffee steaming in a mug by my elbow. Add these extras and it comes to much more than twenty-five pounds. So what do I choose? A pair of corduroy pants. A few thick sweaters. Good boots. A woolen cap. A book with blank pages. A pen. Ottewell's *Astronomical Calendar*, Norton's *Star Atlas*, and Thoreau's *Walden*. I put all of these things on the bathroom scale, and they come to twenty-one pounds. They will do.

Thoreau's friend Emerson once said: "He who knows what sweets and virtues are in the ground, the water, the plants, the heavens, and how to come at these enchantments, is the rich and royal man." Of course, it is self-indulgent to compare my hankering for a less burdened life to the forced poverty of the Kung tribesmen. And, if truth be told, there is little in my four-ton pile that I would be willing to do without. There is no way back to the Kalahari, nor do I have the desire to go. So I will add to my pile sweets and virtues from the sky. I will stack on stars and galaxies until my pile topples over. I will be insatiable. No royal vault will be large enough to hold my riches.

❧

When Galileo turned his telescope on the "Beehive" in Cancer, he counted thirty-six stars. His wealth increased enormously. To the three stars of Orion's belt, he added fifty. When he examined the Milky Way with his instrument, the stars he saw defied enumeration. The treasure he acquired on those few cold nights in January of 1610 could not be contained in all the chests of the Medici.

To Galileo's trove I will add the 500 billion stars of the Milky Way Galaxy and claim them as my own. Let me write that in my ledger: 500,000,000,000 stars. I will conservatively estimate that each star has a half-dozen major planets, several dozen moons, and 10,000 asteroids. That's 5,000,000,000,000,000 potential house plots I own in this neighborhood alone. How about a nice galaxy in the Virgo cluster? There are 3000 visible galaxies in the constellation Virgo, carefully catalogued by astronomers, each with half a trillion stars. I can let you have one cheap.

I hesitate lest I sound like the businessman whom Antoine de Saint-Exupéry's Little Prince met on Asteroid 328, the businessman who owned the stars—or *thought* he owned them. He had his stars all counted out on slips of paper; five hundred one million, six hundred twenty-two thousand, seven hundred thirty-one, to be exact. In de Saint-Exupéry's portrait the businessman is an object of derision, bent on accumulating idle wealth. But I say that a ledger full of stars is not a bad thing. Better than a mansion full of velvet cushions.

The Little Prince lived on a planet scarcely larger than a house, with three volcanoes (two active, one extinct), a rose, and possessions that would certainly weigh less than twenty-five pounds. I envy the Little Prince the size of his tiny planet. One evening just by moving his chair he was able to watch the sun set forty-four times. An evening of forty-four sunsets is worth having. Once I saw three sunsets in a single evening and felt I was a wealthy man. The first occurred as I sat in a plane on the tarmac at Logan Airport in Boston waiting for takeoff. As the plane lifted into the air and headed north I caught a glimpse of the sun over the horizon where it had just set. Then it set again. The plane turned and sped west, chasing the sun. Again I saw the sun rise above the horizon. As the plane turned south toward New York, the sun set a third time. I recorded three sunsets in my ledger.

There are other prized possessions recorded there: the nova that appeared in Cygnus in the autumn of 1975, the extraordinary aurora borealis of September 30, 1961, an eyelash-thin moon no older than thirty hours, the delicately beautiful Comet West dangling by its tail in a blue-pink dawn. There are pages in my ledger for eclipses, occultations, conjunctions, bolides. I acquire these things and hoard them greedily. There are other treasures I would like to own but do not. Many years ago I read an article in *Scientific American* about the green flash, a fleeting strip of color that is sometimes seen near the upper limb of the rising or setting sun. It is caused by atmospheric refraction. To see it you need a perfectly flat and clear horizon, and lucky circumstance. I don't know of anyone who has seen the green flash other than the author of the article. But I have often looked for it. Every time I find myself near a sea horizon and the day is clear, I look for the green flash. I have not yet seen it.

❧

One day I met a friend in a corridor of my college. We both carried large books: hers an anthology of Elizabethan poetry, mine a volume of the *Smithsonian Star Catalog*. "Oh! Stars!" she said. "How wonderful!" I opened my book. It was full of numbers. The coordinates and specifications of 96,000 stars, in a book the size of the New York City telephone directory. Right ascensions and declinations, proper motions, radial velocities, spectral types, visual and absolute magnitudes, distances in parsecs. My friend's bright expression faded. "Grief brought to numbers cannot be so fierce," she said, quoting John Donne.

The stars in my *Smithsonian Catalog* were not the fierce stars my friend hoped to find. They were not the stars the Little Prince gave to his friend the aviator, stars that tinkled like little bells and sang like rusty pulleys. Poets spurn this business of numbering stars. They would prefer the spangled night unquantified. They would prefer to see night's candles burning in a summer sky like tapers in a cathedral. Tapers in a cathedral are well and good, but without the catalogue of numbers we would not have had access to the universe of the galaxies. Without the numbers we would still reside in the compact Earth-centered egg of the Elizabethans. Numbering was the method that let us break the seventh sphere and glimpse the Infinite.

Numbers can be revelations. The sixteenth-century cabalistic text *Zohar* tells us that in every word of Scripture there are many lights. Each word has seventy "facets," corresponding to the seventy known languages of the seventy nations of the Earth. Every separate letter of every word also has seventy facets. For each of these seventy times seventy light-revealing facets in every word there are 600,000 "entrances," or possible interpretations, one for each of the 600,000 witnesses who were present with Moses at Sinai when the Word was first revealed. In every word of Scripture, then, there are 2,940,000,000 lights; on a page of Scripture as many lights as there are stars in the Milky Way. The entire body of the Scripture—the words, the letters, even the shapes of the letters, according to the *Zohar*—reveals the secret name of God. No letter can be altered or removed without tearing a hole in the fabric of the universe.

Modern astronomy is close in spirit to the *Zohar*. The galaxies are the words, and the stars are the letters. Every star has 600,000 entrances, and some are numbers and some are voices that sing with the music of rusty pulleys. No jot or tittle can be discarded without risk. Alpha Centauri is 4.3 light-years away. That's 25,000,000,000,000 miles. You could quarry the Earth and not have sufficient milestones to mark the way from here to there. And every mile is essential to the journey. Number them, name them, the miles and the stars. "He tells the number of the stars, He calls each by name," sings the psalmist. Vega. Capella. Aldebaran. Betelgeuse. But these are only the giant stars. Start closer to home, number them all, every one in order. The sun. Alpha Centauri A, B, and C. Barnard's star. Wolf 359. BD+36°2147. Sirius, and its white dwarf companion. Luyten 726–8.

If I own something like the *Zohar*, it is Robert Burnham's *Celestial Handbook*. Three volumes and 2138 pages of star facts and lore. Turn to any page at random and I find a new entrance to the Infinite. *Page 213:* Nova Aquilae 1918, the most brilliant nova of the past 300 years, first noticed on the night of June 8, 1918, by E. Barnard, of Barnard's Star fame, who was in Wyoming for the purpose of observing an eclipse of the sun that had occurred only a few hours earlier. At the same hour, the new star in Aquila was independently discovered by seventeen-year-old Leslie Peltier of Delphos, Ohio, who later became America's champion discoverer of comets. Within hours of its discovery the new star outshone every star but Sirius. The star that flared up in Aquila was 1200 light-years away and had a luminosity 440,000 times that of the sun. *Page 940:* the planetary nebula NGC 2392, midway between Kappa and Lambda in Gemini, discovered by William Hershel in 1787. According to Burnham, the nebula of fuzzy light has the features of W. C. Fields, but the popular name is "Eskimo nebula," from the funny face ringed in "fur" that appears in a moderate-sized telescope. Careful measurements show that the Eskimo nebula is growing at a rate of seventy miles per second. *Page 1619:* the Milky Way in Sagittarius, the galactic center, hot with stars, blazing with the spectra of formaldehyde, carbon monoxide, the OH radical, methyl alcohol, all the stuff of life. The alcohol content of the star cloud Sagittarius B2, if purged of impurities, would yield 10,000,000,000,000,000,000, 000,000,000 fifths at 200 proof, a quantity of drink that vastly exceeds the entire mass of the Earth.

And here, in the pages given to the star clouds of Sagittarius, in an essay on the staggering number of stars to be counted in that constellation, Burnham tells us of the fourth-century hermit-poet T'ao Ch'ien, who found rather early in life that society attached no particular value to his talents, and that he could no longer "kow-tow for five pecks of rice a day." T'ao Ch'ien retired to his own private world, where he devoted himself to the things he loved: nature, flowers, children, stars, poetry, wine, and his tiny garden with its *three paths* and *five willows*, where on summer evenings he watched the Milky Way rise *over the eastern fence.* And suddenly we are back at Walden Pond, or on the tiny planet of the Little Prince, as poor as church mice and as rich as lords. I count every star in Sagittarius as mine. I kowtow to no one for their possession.

13

THE
BLANDISHMENTS
OF COLOR

In her collection of critical essays *Diana and Nikon*, Janet Malcolm speaks of the difference between black-and-white and color photography. "It is black-and-white photography," she says, "that demands of the photographer close attention to the world of color, while color photography permits him to forget it." Of course, there are magnificent artists in the medium of color film, but it has traditionally been the rejection of color that separates the serious photographer from the snapshooter. The black-and-white medium is hard, says Malcolm, color easy. The former requires art, the latter doesn't.

As I write these lines the trees outside my window are gaudy with color. In New England in October anyone can take a pretty picture, and most do. The cerulean blue sky, the lead white of church spires, the bold brushstrokes of deciduous reds and golds . . . few places in the world can be more beautiful. Point your camera in any direction, snap the shutter, and the image cannot fail to please. Color does it for you. The eye gets lazy.

In a few weeks November comes and the leaves will be gone. Then begins the season when the eye must pay close attention, the season of black and white. Color in November is caught in brief, evasive glimpses. The tiny crimson berry of the Canada mayflower. The cap of the golden-crowned kinglet glittering in the branches of the spruce. A speckle of pink feldspar in the granite outcrop. The eye works hard for the color it finds in November.

The serious photographer, says Malcolm, resists "the blandishments of color." Walking the woods in November is like doing black-and-white photography. It is the eye that must compose nature's beauty. It is the eye that must frame the pale rose breast of the nuthatch against the crusty black skin of the pine. It is the eye's subtle chemistry that fixes in the curled leaves of the winter beech the color of freshly baked bread. "The beech is a bakeshop open all winter," writes the poet Maxine Kumin, and art has made of slight color a masterful image.

But today, ostentatious October fills my window, demanding attention, squeezing its energy into every inch of the frame. In a month, all of that technicolor razzle-dazzle will be gone. October's color redounds to nature's credit. Color in November is the work of art.

❧

Like color in November, color in the night sky is the work of art. The night sky is November all year round. Night is the black-and-white print. The stars twinkle in a mono-chrome of pure silver chloride. The stars are resonantly colorless.

Or are they? A little artfulness will show that the stars are not mere points of white light. Many stars are in-deed white. But others have tints of red, orange, yellow, or blue. Some, it is said, are green or purple. The reason the stars appear white to the impatient eye is an accident of the eye's chemistry.

There are two kinds of light receptors on the retina of the eye: the rods and the cones. The cones are the color sensors, but they do not respond to faint illumination. The rods are more finely attuned to dim light, but they do not discriminate colors. When we look at the stars, it is the sensitive but color-blind rods that do most of the work of seeing, and that is why the stars appear mostly white. But turn a telescope toward the stars, or take the time to look at the brighter stars with care, and they will take on the colors of the rainbow.

The color of a star is determined by the temperature of its surface. In this respect, a star is like any other incan-descent object. An object just hot enough to reach incan-descence glows with a dull red light. As the temperature goes up, the amount of energy radiated at the shorter (bluer) wavelengths increases relative to the longer (redder) wavelengths. As the object grows hotter, its appearance changes from red, to orange, to yellow, to white, and fi-nally to blue. Or perhaps I should say from reddish, to or-angish, to yellowish, to white, to bluish-white, for the colors of the stars are never pure but a mixture of all wavelengths; it is the dominance of a certain part of the spectrum that determines the color we perceive. The surface temperature of our sun is about 5800 degrees Centigrade, and like any incandescent object of that temperature it radiates a predominantly yellow light. No child fails to put a yellow face on the sun. The sun's yellow hue is apparent to the eye because that star is near and bright, and the cones of the retina go about their business of "seeing" color with efficiency. But put the sun at the distance of Alpha Centauri A, a yellow star that is the sun's exact twin, and you will see only a white dot on a black sky.

A careful observer can distinguish the colors of the brighter stars in the night sky. Often I have pointed to Antares and said "a red star," or to Vega and said "blue." That's a bit of an exaggeration. Antares is classified as a red giant star, but it will probably appear pale orange to the naked eye. Vega, if it looks blue at all, is white with a bluish cast. And the sun, in spite of the child's yellow face, is really more white than yellow. Perhaps the best way to see star colors is by contrast. Orion offers a vivid demonstration. Rigel, the star in the giant's forward foot, is decidedly blue. Betelgeuse, in the raised arm, is orange. The colors are most easily seen by glancing back and forth between the two stars. Star colors are more readily perceived if the intensity of the starlight is amplified with binoculars or a telescope. My favorite demonstration of star color is little Albireo, the star at the beak of Cygnus the Swan. Albireo is a binary star and appears double in a small instrument. One member of the pair is golden, the other a brilliant blue. Once you have seen the blue and gold of Albireo you will never again think of the stars as uniformly white.

The art of observing the night sky is 50 percent vision and 50 percent imagination. Nothing illustrates the truth of the saying better than the colors of the stars. Nineteenth-century observers seem to have had the best luck at seeing the colors of stars, no doubt because their observations included a larger dose of imagination relative to vision. Richard Hinckley Allen's book *Star Names: Their Lore and Meaning,* which came out just at the end of the century, calls Albireo "topaz yellow and sapphire blue." Antares, also double in a fair-sized telescope, is referred to as "fiery red and emerald green." Allen describes other stars as *straw, rose, grape,* and *lilac,* and you would think you were in his garden rather than the sky. Allen's color descriptions were mostly borrowed from the celebrated British observer William Henry Smyth, whose eye was apparently refined enough to see a dozen shades of white, including *pearly, lucid, creamy, silvery,* and just plain *whitely white.*

The most elaborate nineteenth-century system of star colors was that of the Russian–German astronomer Wilhelm Struve. Struve's classification used Latin labels, and the words themselves have an exotic ring, like the Latin names of tropical birds. *Egregie albae, albaesubflavae, aureae, rubrae, caeruleae, virides, purpureae.* Sometimes these basic descriptions did not seem sufficiently accurate, so Struve invented new ones, like *olivaceasubrubicunda* for pinkish-olive. Show me a star that is *olivaceasubrubicunda* and I'll show you imagination making art of thin stimulus.

Armed with cameras and electronics, the modern astronomer has reduced star color to objective number. Rigel in Orion has a color index of -0.04 and a spectral type of B8, and that says it all to the professional. But the amateur stargazer needs only a little imagination to see that Rigel is blue and Betelgeuse is orange. With Smyth's level of imagination he might even see the "pale rose" of Aldebaran or the "golden yellow" of Arcturus. Allen's book describes the stars of binary Regulus as "flushed white and ultramarine." These are colors no camera can catch.

Stargazing, like black-and-white photography, demands close attention to color. There are no ravishing sunsets in the midnight sky, no deciduous riots of red and gold in the forest of the night. The snapshooter turns from the telescope in despair, but the artful observer will take the hint and let his imagination enrich the palette. William Henry Smyth fixed his telescope on the stars and saw "crocus," "damson," "sardonyx," and "smalt." This is the kind of imagination that labels paint chips. Smyth's descriptions of star colors remind me of the experience of the artist Vasili Kandinsky when he bought his first box of tubed pigments at the age of thirteen. Kandinsky tells us how at the slightest pressure of his fingers on the open tubes the colors slipped out like animate beings, some cheerful and jubilant, others meditative and dreamy. Some colors seemed to emerge "self-absorbed." Others slid from the tubes with "bubbling roguishness," some with a "sigh of relief," still others with a "deep sound of sorrow." Some of Kandinsky's colors were "obstinate," others "soft" and "resilient." The artist's almost mystical experience with colored oils evokes stars of damson and smalt.

It is just as well that the colors of the stars are not easily seen. As John Burroughs said, if the deep night were revealed to us in all its naked grandeur, it would perhaps be more than we could bear. But half of infinity is still infinite. A hint of infinity is infinite too—if we can take the hint. A trait here, a trait there, Burroughs said. Of hints and traits we make our way.

On a night that is perfectly dark and perfectly clear, the naked eye can just discern the Great Orion Nebula as a patch of fuzzy white light in the sword of Orion. In a moderate-sized instrument, the nebula glows with an eerie green light, and the eye sees hints of shape—a shape like a curled hand with the palm aflame. Long time-exposure photographs of the Orion nebula reveal a stunning complex of stars in the trauma of birth, swaddled in vortices and streamers of collapsing luminous gas, a star-cradle measured by light-years and charged with the energy of Creation. The colors of the nebula are as various as the photographs. On Kodacolor 400 film Orion's great cloud is plum and lilac, cerulean blue and milky white. On Ektachrome 400 the nebula is apricot and red, tinted with deep ochers and browns. On Kodacolor 1000 film Orion's nebula is mauve and amethyst, blushed with rose. But all of these colors are artifacts of the film; they are not what the eye would see if we could approach the nebula and stand in its awesome light.

Here on my desk is a photograph of the Great Orion Nebula that was made with a special care to reproduce the color sensitivity of the human eye. The colors of the print were created with three black-and-white time exposures taken in twilight with the 3.9-meter Anglo-Australian telescope. The plates were made with a combination of emulsions and filters that gave a uniform response across the entire visual spectrum. Masking techiques were used to accentuate the delicate structure of the nebula. The Orion nebula is more than 20,000 times larger than our solar system. There is enough hydrogen, helium, and other materials in the cloud to form 10,000 stars like our sun. The light of the nebula comes mostly from the radiation of doubly ionized oxygen (green) and the alpha radiation of ionized hydrogen (deep red). The gas is made to glow by the energy of hot young stars embedded in the nebula, stars only recently born of the stuff of the nebula itself. The Anglo-Australian photograph shows the nebula feathered with filaments of gas like a great bird of prey. In the bird's knarry talons are gripped a dozen intensely white stars. The green oxygen light available in the moderate-sized instrument unites with the reds and blues of the traditional Kodacolor photographs to brown and yellow the light. There are olives and khakis in the composite print, and the reds and oranges of a log fire. There are ominous browns and grays, banked against the black of space like ash and cinder. There is movement and violence; the nebula seems charged with a terrible malevolent power. It is not a pretty picture. This carefully contrived photograph is not a drugstore snapshot of a pretty Vermont village steeped in the colors of October. It is the face of Leviathan, wrenching us into a space as deep and as terrible as the bowels of the sea. It is God's sturdy hand, the fist that grips us in its clinched infinities. This is the power that hides in the colorless night like the rocks in foaming breakers that crack a ship, or the white whale that drags all who seek him to black oblivion.

❦

It would be impossible to end this meditation on the blandishments of color without rereading Melville's chapter on the whiteness of Moby Dick. Ishmael rehearses for us the way in which white stands for what is good and true. He parades for our attention snow-white chargers, the ermines of Justice, the white-robed four-and-twenty Elders before the great white Throne, and Jove in the guise of the snow-white bull. But all of these he dismisses when he turns to Moby Dick. It is the whiteness of the whale, he says, that above all else appalls him. In spite of the associations of whiteness with "whatever is sweet and honorable, and sublime," there yet lurks an elusive something in the innermost idea of white that strikes "more of panic to the soul than that redness which affrights the blood." When coupled with any object terrible in itself, concludes Ishmael, whiteness heightens the terror to the furthest bounds.

If the retina of the human eye had only rods, and those with the sensitivity of cones, then the stars would shine with October's reds and golds, and Orion's nebula would unfold like a flower of ocher and green at the hunter's knee. If that had been so, then perhaps we would never have sought divinities in the sky. Then perhaps ours would have been a snapshooter's theology, a philosophy of the drugstore print, a metaphysics all prettified and simple. But the night, like Ahab's whale, cloaks its immensity in whiteness. *The great principle of light touches all objects in the night sky, stars and nebulae, with its own blank tinge, and the palsied universe lies before us like a leper* (the words are Ishmael's). We gaze ourself blind at the monumental shroud that wraps our brightly colored planet. The pilgrim who would find his way to the edge of the galaxies and to the beginning of time must forgo daylight's easy color and launch himself upon the black-and-white sea of the night and in those huge spaces find stars the colors of damson, crocus, grape, and straw. The quest will perhaps require more courage than you or I can bear, and there is the possibility that we will be drawn, like Ahab, into the starry deep, lashed to the object of our search with the lines of the chase. "Of all these things," says Ishmael, "the Albino whale was the symbol. Wonder ye then at the fiery hunt?"

14

FOLLOWER
OF THE PLEIADES

W ho can tell the name of the huge red star now rising in the east? Or of the bird last summer that made haunting cries in the night and flapped up from its hiding place like a large brown moth? Nowadays there is little incentive to learn the names of things. The nominalist philosophies on which we rest our ways of knowing emphasize the arbitrariness of labels. A name is an empty x to be assigned at will. Stout little x staggers through our textbooks carrying all the burdens of knowledge. Now it is the celestial coordinates of that red star in the east (4h 34.8m R.A., +16°28' dec.), now it is the Linnean designation of the bird that cries in the summer night *(Caprimulgus vociferus).*

Thoreau tells us that when he learned the Indian names for things be began to see them in a new way. When he asked his Indian guide in Maine why a certain still lake was called Sebamook, the guide replied: "Like as here is a place, and there is a place, and you take water from there and fill this, and it stays here; that is Sebamook." Thoreau compiled a glossary of Indian names and their meanings. It was like a map of the Maine woods. It was a natural history. The Indian names of things reminded Thoreau that intelligence flows in channels other than our own.

No part of our environment is so rich an archive of other intelligences as the night sky. The night is a repository of human cultural history. The names of the stars are entries in a family album that show us what we have been and what we have become. Some star names are adjectives that describe the stars: Sirius, for example, means "sparkling" or "scorching." Arcturus takes its name from its place in the sky, not far from Ursa Major; it is "the guardian of the Bear." Some names refer to the place of the star within a constellation: Betelgeuse, of Arabic origin, probably means "the hand" of Orion. Most of our star names are Arabic: Zubenelgenubi and Zubeneschamali, the "southern claw" and "northern claw" of the Scorpion (now part of Libra), are among the more exotic examples of Arabic names. Greek and Latin names are not uncommon among the stars; Canopus is from the Greek, Capella is Latin.

Nunki in Sagittarius is Sumerian; and at least one star name on Western maps, Tsih in Cassiopeia, made its way from China. Nunki harkens back to prehistory; Cor Caroli, "the heart of Charles" (II of England), is a late arrival—Newton's colleague Edmund Halley named the star to honor his monarch. The star names Sualocin and Rotanev, in Delphinus, sneaked onto sky maps when Piazzi, in his *Palermo Catalogue* of 1814, reversed the names of his assistant, Nicolaus Venator, and attached them to stars, confounding later etymologists who puzzled over their derivation.

It is said that Adam had the privilege of naming everything in Paradise. I would like to imagine that he took that responsibility seriously. He sat beneath the Tree of Knowledge and thought and thought, and then with a sudden insight he leapt to his feet and whispered, "Willow." And willow it was and willow was exactly right and willow it has been ever since. And it occurred to him to think "willowy Eve" and "weeping willow," and language took flame on his tongue and the flame spread from tree to bush, from bush to bird, from bird to star, and soon all of the night was ablaze with parts of speech. *Scorcher, guardian, heart of Charles:* the stars burn in our intelligence.

❦

The bird in the summer night that flapped up from hiding like a large brown moth cried its own name. It was the whippoorwill. And the red star that even now is rising outside my window is Aldebaran. Let me tell you about that star. Aldebaran is the brightest star in the constellation Taurus, and so it earns for its stout x the designation Alpha Tauri. In the *Henry Draper Catalog* it is listed as 29139, and in the *Smithsonian Astrophysical Observatory Catalog* it is 94027. We see Aldebaran in the sky as one of the cluster of stars called the Hyades, in the face of the Bull. From time immemorial Aldebaran has reigned among the Hyades like a queen bee in her hive. But Aldebaran is not properly a member of the cluster—it is only half as far away as the stars of the Hyades. The distance to Aldebaran has been measured by direct triangulation (the so-called method of parallax), the way a surveyor would measure the distance to a remote peak by laying out a baseline and measuring angles. When astronomers measure the distance to stars they use the diameter of the Earth's orbit as a baseline. But even with so large a base, only the nearest stars are close enough for the method to work. Aldebaran is barely within range of triangulation. It is sixty-eight light-years away. If you set out to drive to Aldebaran, minding the fifty-five-miles-per-hour speed limit, it would take you 800 million years to get there.

Aldebaran is a star of spectral class K5, which means it has a surface temperature of 4000 degrees Kelvin, and even though I have called it red (the Hindus called it the "red deer"), it more accurately appears pale orange to the naked eye. Aldebaran is a giant star (but not a supergiant like Betelgeuse or Antares), a star approaching the end of its life, puffed up and dying. It may be slightly variable, as dying stars often are, its surface heaving slowly in and out with a kind of labored breath or sighing. Some tens of millions of years ago Aldebaran was a mid-sized yellow star like our sun, perhaps modestly warming a comfortable family of planets. The planets of Aldebaran, if it had any, have now been seared by the huge heat of that star's terminal trauma.

Like other stars, Aldebaran moves across our sky, although not enough from night to night, or even in a lifetime, for us to notice the displacement. It was, in fact, one of the first stars whose motion in the sky was discovered. An occultation of Aldebaran by the moon was recorded in Athens in March of A.D. 509; on that night the moon passed directly between the Earth and the star, briefly blocking the star's light. Aldebaran is one of the few first-magnitude stars that lie close enough to the moon's apparent path in the sky for this to sometimes happen. Several times I have myself watched occultations of Aldebaran. On one of those occasions, it was the unlit edge of a crescent moon that approached the star; one instant the star was shining half a degree from the crescent moon, and then—click!—it was gone, like a coin in the hand of a prestidigitator. Newton's friend Edmund Halley recognized that the moon could not have occulted Aldebaran in March of the year 509 unless at that time the star was a fraction of a degree farther north, and so he concluded that the star had changed its position on the celestial sphere. We now call this slow slide across the bowl of night the *proper motion* of a star, and here in my catalogue it says that Aldebaran has a proper motion of 0.2 seconds of arc per year, which means that in the nearly 1500 years since the Athenian occultation Aldebaran has moved a distance equal to a fourth of the moon's diameter. In addition to this slow lateral slide in the direction of the constellation Orion, Aldebaran is moving away from us, a motion that can be detected by a slight stretching or reddening of the star's light. In the 1500 years since the occultation at Athens, the distance between us and Aldebaran has increased by 2 trillion miles; if you were driving to Aldebaran at fifty-five miles per hour, you would never get there at all.

So now I have paraded Aldebaran's x's, and it is out of these x's that we say we know that star, and it is out of the accumulated x's of Aldebaran and all of the other stars that we have come to know that the universe is not the star-studded body of a goddess or the many-storied mountain that Dante climbed with Beatrice. It is out of the accumulated x's—the right ascensions and declinations, parallaxes, proper motions, radial velocities, magnitudes and spectral types—that we have discovered the universe of the galaxies and quasars, the universe of light-years and infinities. Flexible x, the empty vessel, the chameleon, has been our instrument of knowing.

Aldebaran is the star's name, even to the astronomer who on the necessary occasions notes it down as HD 29139. Aldebaran means "the follower," from the Arabic *Al-Da-baran*. Most Arabic star names are translations of earlier designations of Greek or Hellenic origin, but *Al-Dabaran* was in use in Arabia before that people had any contact with classical science. The Arabs of old used the stars to distinguish the seasons and to navigate featureless desert seas. The names of the stars were as well known to the simple desert sailor as to the astronomers and mathematicians of Alexandria.

Aldebaran "follows" the Pleiades, that oasis of faint stars in the empty quarter of Taurus. The six or seven naked-eye stars of the Pleiades are not particularly bright, but there is nothing else like them in the sky. When the Pleiades appear above the eastern horizon you can be sure that Aldebaran will rise an hour later at the same place and follow that glittering cluster across the bowl of night.

Other Arabic names for the star can be translated as "fat camel" or "driver of the Pleiades." Aldebaran's ancient Roman name, *Palilicium*, commemorated the Feast of Pales, the special deity of shepherds; and shepherds watching flocks by night knew the star's rising. Ptolemy called it "the Torch Bearer," and to the Babylonians it was *I-ku-u*, the "leading star of stars." To most modern watchers of the constellations, Aldebaran can be nothing else but the fierce red eye of the bull that charges at Orion.

Follower, driver, fat camel, red eye: Names illuminate the sky like an aurora; they enfold the stars in curtains of intelligence. "The most visible joy," says the poet Rainer Maria Rilke, "can only reveal itself to us when we've transformed it, within." Rilke gives his own names to the stars and constellations, animating and transforming them. "There, look," he cries, "the *Rider*, the *Staff*, and that fuller constellation they call *Fruitgarland*. Then, further, toward the Pole: *Cradle*, *Way*, *The Burning Book*, *Doll*, *Window*." The "animated, experienced things that share our lives" are passing away, claims Rilke, crowded out by pseudo-things, things "empty and indifferent." The usurper, presumably, is little x, marching in ranks with its brothers, an irresistible infiltration.

But surely in this matter Rilke is wrong, for x has itself been an effective instrument of transformation. Science makes good use of x, uses it like a beast of burden to bear an enterprise that has taken us beyond the cold desert night of the bedouin, beyond the fat red star that tends the Pleiades as a shepherd minds his goats, into a universe more wonderful than any shepherd might have dreamed. In science, little x is an appliance, a tool that lets us tinker in a mathematical sort of way with the infinite machinery of the universe. It is an immensely effective tool because the machinery of the universe seems to be in some mysterious and remarkable way mathematical. Because of x, we live in a night hung with red giant stars that swell and burn with a red thermonuclear light, and not in a night of fat camels or red deer. Intelligence has turned to flow in new channels. I will follow that stream. I will transform those red giant stars. I will animate them. I will make those huge red stars my own.

❧

It is a continuous story. Ptolemy of Alexandria referred to Aldebaran as "The Torch Bearer," and the Romans made that star commemorate the Feast of Pales. When classicial science waned after the fall of Rome, astronomy and the sciences generally died out in Europe. But the Arabs kept the old traditions alive; they translated Greek astronomical lore into the language of Islam, and they infused Greek astronomy with words of their own. It was from Arabic sources that Europeans recovered astronomy in the Middle Ages and the Renaissance. Spain became the most important center of this transmission, for there Arabic and Christian scholars had mingled since the Islamic con-

quests of A.D. 711. In the Christian parts of Spain translations were made of the Arabic texts from the late tenth
century onward; it was from these texts that we received
the name of the red star in Taurus, transformed into something resembling Latin. "The Follower" retained its Arabic
name in Europe until 1603, when Johann Bayer, in the
spirit of the scientific revolution, imposed Greek letters on
the stars in a more or less systematic way, and "the Follower" became Alpha Tauri. Aldebaran became HD 29139
in the *Henry Draper Catalog*, compiled by Annie Jump
Cannon and E. C. Pickering between 1918 and 1924, and
SAO 94027 in the *Smithsonian Astrophysical Observatory
Catalog* of 1966. And so the stream of intelligence has
twisted and tumbled over the uneven topography of cul-
ture, now finding one channel, now another, always flowing toward the distant ocean that is truth.

The night will not be empty or indifferent as long as
we mind the stars, naming them—"fat camel," "Alpha
Tauri," "HD 29139"—transforming them within. I repeat
the names of the stars in a kind of litany: Aldebaran, El
Nath, Rigel, Betelgeuse, Bellatrix, Alpha Centauri, Barnard's Star, Wolf 359, BD + 36°2147 (with its suspected un-
named companion). It is we ourselves who give the stars
their invisible reality, beyond the visible. By watching. By
naming. They depend upon us, says Rilke; we are their
transformers, "our whole existence, the flights and plunges
of our love, all fit us for this task."

15

THE SHAPE
OF NIGHT

Z en teaches that all things, whether consciously or unconsciously, are in search of their own true nature, their buddha nature. A buddha is one who has awakened to his real self; he is "at home in a homeless home." Among the many buddhas there are sun-faced buddhas and moon-faced buddhas. Sun-faced buddhas have lived a long time in the light of the truth about themselves, moon-faced buddhas only briefly.

I am a cautious pilgrim of the night, a tentative wanderer among the stars. My awareness of my home in the universe is fleeting and incomplete. Into the homeless home of the sun-faced buddha I have stepped but briefly. My quest, such as it is, is rewarded with faint lights and scrawny cries, a trait here and a trait there, a hint of the infinite and a tingle in the spine. Of "minute particulars" I will make my way. Give me, then, the moon face. Give me the blood-red face of the moon.

❦

In May of 1503, on his fourth voyage to the New World, after many trials and adventures, Christopher Columbus sailed with two ships from Panama, intending to stop at Hispaniola for refitting before returning home to Spain. Crippled by storm and riddled by worms, the little fleet was run ashore on the north coast of Jamaica. Columbus sent twelve men in canoes to seek rescue from Hispaniola, 200 miles to the east. Then, for months, he waited with the remainder of his men for rescue. To obtain food, the Spaniards bartered beads and mirrors with the local Indians. Eventually, the natives tired of trinkets and balked at providing provisions for the stranded sailors. Columbus saw a solution to the problem. He had with him a copy of Regiomontanus' *Ephemerides*, which contained a prediction for an eclipse of the moon at moonrise on the night of February 29, 1504. Columbus called a meeting of the local chiefs and declared that if food were not forthcoming he would cause the moon to rise "inflamed with wrath." And, as he said, even as the moon rose it was the color of blood.

On the leap-year night of February 29, 1504, the moon rose just at sunset and slipped into the shadow of the Earth. Unlike other shadows, the shadow of the Earth is red. It is stained the color of blood by long-wavelength sunlight refracted by the atmosphere around the curve of the Earth. As Columbus and the Indians watched, the Earth's red-stained shadow moved across the face of the full moon and the moon was transformed from a gold doubloon to a dusky disk of crimson. I have on many occasions watched eclipses of the moon. The effect is spooky, mysterious. If I had not known about the cone of night I would have been as chastened as the Indians of Jamaica.

❦

Night has a shape and that shape is a cone.

In Shelley's *Prometheus Unbound* the Earth speaks this line: "I spin beneath my pyramid of night, which points into the heavens—dreaming delight." When I first read that line many years ago I was startled by the recognition of something I had possessed all along. I had studied astronomy and optics. I knew about umbras and penumbras and the way objects cast shadows in different kinds of light. In my astronomy classes, I had drawn the necessary triangles to calculate the relative sizes and distances of the sun and Earth and moon. I suppose I had *known* all along that the Earth's shadow is cone-shaped and points darkly into sunlit space. But until I read Shelley's line I had never *experienced* night as a tall pyramid of darkness receding from the globe.

Earth wears night like a wizard's cap. The wizard's cap is long and slim and points away from the sun. It is 8000 miles in diameter at the rim, where it fits snugly on the Earth's brow. It extends to a vertex 860,000 miles from the Earth. The wizard's cap of shadow is a hundred times taller than it is wide at the base. It reaches out three times farther from the Earth than the distance to the orbit of the moon, and when the moon in its monthly circuit happens to pass through that cap of darkness, we have an eclipse of the moon.

When next I watched the moon eclipsed it was Shelley I recalled, not the astronomy text. The curved surface of the Earth's pyramidal shadow moved across the moon's full face and traced cone-shaped night. Night has not been the same since. Now night has a shape. It is the difference between *knowing* and *seeing.* There is a Zen story about a man who when he was young saw the trees as trees, the wind as wind, and the moon as the moon. As he grew older he began to ask himself why the trees grew as they did, why the winds blew from the four corners of the Earth, and why the moon waxed and waned. Everything he saw posed a question, and all of his time was spent pursuing answers. Then there came a time when the trees were again trees and the wind was again the wind and the moon was again the moon. After I read Shelley, I watched an eclipse of the moon. The moon slipped into the Earth's red shadow. For a moment the moon had the round red face of the buddha, poised in the pyramid of night. For a moment it was just the moon.

❦

The Earth spins beneath its cone of night. The Earth orbits the sun, and its ruddy shadow cap goes with it, always pointing toward infinity. Under that darkling cap badgers scuttle in ditches, stalking night-crawling slugs and beetles. Bats thrash the air and squeal a high-pitched cry that only children hear. Owls in oaks hoot at the moon. Under that darkling cap go possums, foxes, raccoons, the creatures with the big eyes, glowworms, friar's lanterns, will-o'-the-wisps. Under that darkling cap go spooks and goblins, incubi and succubi, bogeys and banshees, the Prince of Darkness. Astronomers mount their tall chairs and point their instruments into the long cap, following Earth's shadow up the Chain of Being, level by level, choir by choir, rank by rank, past the Fortunate Isles, through Elysian Fields, beyond Zion, and into the sea without shore where stars and galaxies beckon and quasars frighten like St. Elmo's fires.

Night is a cone because the Earth is round and smaller than the sun. It may have been the observation of the rounded curve of the Earth's shadow on the moon during an eclipse that led the Greeks to the astonishing discovery that the Earth is a sphere. Of course, they first had to have guessed that it was *the Earth's shadow* that caused the eclipse of the moon, rather than sky dragons or malevolent gods. The Greeks were mathematical. The Greeks replaced Baal and Zeus with Euclid. They drew circles and straight lines in the dust and saw in those constructions that the night is a cone. The straight edge and the compass were their telescope. But it took more than mathematics for the Greeks to *experience* the spherical Earth. In the end, it was *imagination* that gave the Earth its spherical shape, an act of insight as pure and focused as a circle is pure and focused, an insight that flashed straight away to the truth like a straight line. I have a friend, Mike Horne, who teaches astronomy with that kind of imagination. I have often watched him with groups of students under the stars, cavorting outrageously to convey a feeling for the Earth's sphericity. He stretches his long arm to point to the sun somewhere over the horizon in Asia's sky. He arches his eyebrows and stands on tiptoes to peer over the curve of the Earth for a star that set the previous night. He moves his hands in great round circles as if to embrace the globe. As I watch him, I find myself pushing up onto tiptoes and lifting my eyebrows, sympathetically, the way a parent opens his mouth when feeding a child, to encompass in my mind's eye the huge round hump of the Earth.

All of the planets wear caps of night. Every object

near a star casts a pyramidal shadow. If an astronaut floating free in Earth orbit puts his feet toward the sun, his wizard's cap of darkness is a hundred feet long. The moon's
shadow cap, by a wonderful coincidence, is almost exactly
as long as the average distance of the moon from the Earth.
If the moon is near apogee (its greatest distance from the
Earth) and passes exactly between the Earth and the sun,
the vertex of its shadow falls just short of the surface of the
Earth. In this circumstance, the sun is still visible around
the moon as a ring of brilliant light, and we have what is
called an annular (ring-shaped) eclipse of the sun. When
the moon is not near apogee its shadow reaches to the
Earth, and when the moon passes between the Earth and
sun the apex of its shadow slices into the Earth like a surgeon's knife. Those who are lucky enough to live within
the swath of that stroke will experience one of nature's
most spectacular special effects, a total eclipse of the sun.
Viewers of a total eclipse stand in the tip of moon's night,
a few moments of borrowed darkness in daytime when
bats fly, owls hoot, and badgers peek from their burrows.

Sometimes, when the moon is just the right distance
from the Earth, its shadow brushes the surface of the Earth
as gently as the tip of a feather, and the eclipse of the sun
is at the borderline between annular and total. The eclipse
of May 30, 1984, was just such an eclipse. The moon's
shadow reached to within such a narrow distance of the
Earth's surface that one could almost have jumped into its
vertex. The tip of the dark feather passed just above New
Orleans, Atlanta, Greensboro, and Petersburg, Virginia,
and moved out to sea along Maryland's eastern shore. The
highest mountains of the moon's limb reached to the sun's
edge and covered it. But sunlight glimmered through lunar
valleys like diamonds on a necklace, like a bowl of light
reduced to its broken glittering rim. In Shelley's poem, this
is the full line the Earth speaks: "I spin beneath my pyra-

mid of night, Which points into the heavens—dreaming delight, Murmuring victorious joy in my enchanted sleep; As a youth lulled in love-dreams faintly sighing, Under the shadow of his beauty lying, Which round his rest a watch of light and warmth doth keep." And the Moon responds: "As in the soft and sweet eclipse, When soul meets soul on lover's lips." On May 30, 1984, the moon interrupted the sun's watch of light and warmth with a kiss so gentle it was barely felt.

❧

Every object near a star wears a cone of night. Near every star there is a ring of cone-shaped shadows that point into space like a crown of thorns. The sun's family of nights includes the shadows of nine planets, several dozen moons, and an army of asteroids. Every particle of dust in the space of the solar system casts its own tiny pyramid of darkness. The sun bristles with nights like a sea urchin prickly with shadowy spines.

Earth's cone of night is the Paraclete that brings the gift of deep space and deep time. On the planet's daylight side the atmosphere scatters sunlight into an obscuring blanket of blue, the Earth's "blue Mundane shell," William Blake called it, a "hard coating of matter that separates us from Eternity." But when we turn with the spinning Earth into night's dark cone, we glimpse the universe. Years ago I read a science fiction story about a planet in a system with four suns. Only once in 2000 years did all four suns set at the same time and the sky go dark. When that singular event came to pass, the people of the planet saw night for the first time and were overwhelmed by its majesty.

The pyramid of night is Earth's narrow chink in its blue armor. Blake says: "If the doors of perception were cleansed, everything would appear to man as it is, infinite. For man has closed himself up, till he sees all things thro' narrow chinks of his cavern." The blue air closes us up. Only through the crack of night do we glimpse the Infinite. Only through the crack of night can we seek our sun-faced buddhas. Through that cone-shaped chink in the Earth's blue mundane shell we court Infinity the way Pyramus courted Thisbe.

16

A MIDWINTER
NIGHT'S DREAM

T his morning I was up and out an hour before the dawn. The air was cool and still, the sky perfectly clear. I walked through the woods to a place where I would have a long, clear view across the meadow to the east. I knew what I would see. This morning promised another rare chance to see all five naked-eye planets in the same part of the sky at the same time.

High in the south, Mars and Saturn were bracketed between the stars Antares and Spica. Mars mimicked Antares' ruddy glow, and Saturn copied Spica's white brilliance. Above the trees at the far side of the meadow, Venus and Jupiter were in conjunction. Venus and Jupiter are the two brightest starlike objects in the sky; to see them so close together would have been reason enough to wake early and come to that silent place. I knew from my ephemeris that Neptune and Uranus were also in the sky near the place of conjunction of the two bright planets, but only a telescope would reveal them, and I had come to the meadow without an instrument. Pluto, too, was somewhere nearby, but I would have had scant chance of seeing it even with a scope. Between Mars and Saturn, on the one hand, and Venus and Jupiter, on the other, was the crescent moon, slipping, even as I watched, closer to the sun. Now I waited and studied the sky near the place on the horizon where the sun would rise. Already the sky was brightening, and layers of violet, blue, and pink light peeled away like onion skins from the still-hidden sun. Then at last I saw it, little Mercury, a pinprick of light, a mote of dust in the gathering day. Five planets. Five planets strung like beads on the string of the ecliptic. Five planets and the moon. Five planets, the moon, and two brilliant stars arrayed along the zodiac.

Medieval astronomers believed that the universe was a nest of crystalline spheres, concentric with the Earth, like one of those many-shelled Russian eggs with a little doll at the center. This morning it was easy to imagine the shells of that glittering cosmic egg, layer upon layer, receding from the Earth. First, the sphere of the moon, turning perceptibly eastward even as I watched. Then the sphere of Mercury, wobbling back and forth to the tune of the sun. Then Venus, the sun, Mars, and Jupiter, in ascending order, heavenward, each sphere rolled round and round by its own special angelic choir. Finally, enclosing all, the sphere of the fixed stars—Antares and Spica, Deneb and Vega and Altair—the glass container of creation, the fortune-teller's crystal ball. What a tidy cosmos it seemed, like one of those little glass globes we had as children, filled with mineral oil and white flakes; when you tipped them briefly, snow fell on scenes of tranquil beauty—a white-spired church, children making snowmen, a tiny Bethlehem.

❦

The moon in this morning's sky was four days from new. "How slow / This old moon wanes! she lingers my desires," says Theseus, Duke of Athens, in *A Midsummer Night's Dream*. And so the play begins. "Four days will quickly steep themselves in night; / Four nights will quickly dream away the time," Hippolyta responds. In Shakespeare's time the cosmos was still an egg of concentric spheres, layered according to value from inside out; ponderous earth at the center, then water, air, and fire; a hierarchy of plants and animals, ascending from the lowest lichen to man; the ranks of the body politic, peasants, gentry, lords, and king; shells of ethereal substance, moon, sun, and planets; choirs of angels, pure spirits, reaching at last to the foot of God's throne. The moon's sphere was the boundary between the terrestrial world of change—lust, love, greed, corruption, inconstancy, trust—and the immutable world above. See how the moon waxes and wanes; see the planets and the stars on their unyielding paths. Tomorrow and the next day and the day after, the moon will wither. She will slip past Venus and Jupiter, nudge Mercury aside, to meet the sun. Then she will reappear four days hence in the evening sky, "like a silver bow / New-bent in heaven." The massed elegance of this morning's planetary alignment will be gone. We will be on our own then, the moon and I. No early rising. I will feast on evening fare, and the moon will fatten toward full.

Not long ago I read Thoreau's little-known essay on moonlight. A short poetic piece, full of dusk and equivocations. This was his theme: The light of the moon is "not disproportionate to the inner light we have." If all the world's a stage, then the moon yields light enough to illuminate the comedy we play out here below its sphere. Night by night we watch the moon wax and wane. We feel her tides in our thoughts. In sleepless midnight hours we sense the moon's gentle gravity tugging dark thoughts from deep sloughs untouched by daylight. Gone are daylight's grand vertical confidences. On our backs in moonlight we are the elements jumbled. The world is out of joint. Stones rise into the sphere of air, fire sinks, the waters overlap the land, air bursts into flame. The planet is a ship of fools. The candlemaker pretends to be king. The king is belled like a cat. Night is the play within the play, and the moon presides. Out of our subconsciousness come Quince, Snug, Bottom, Flute, Snout, and Starveling, *not impaired but all disordered*. The moonlight is sufficient for their revels.

❦

In the winter of 1609–1610, perhaps even as actors at the Globe Theatre in London played *A Midsummer Night's Dream* with the author in attendance, Galileo in Padua turned his new telescope on the moon. He saw (or thought he saw) dark seas and bright lands, oceans and continents, mountain peaks swelling in morning light, valleys lapsed in shadow. His casual comparison of the moon and Earth became the source of his later difficulty with the Church. Since the time of the Greeks, the idea that the moon was of the same nature as the Earth had been dangerous to hold, disruptive as it was of the cosmic hierarchy that affirmed man's proper place in the universe. Galileo did not shrink from blasphemy. The Earth, he said, "must not be excluded from the dancing whirl of stars. We shall prove the earth to be a wandering body surpassing the moon in splendor, and not the sink of all dull refuse." When Galileo put down his telescope, the cosmic egg had been broken, the nested crystalline spheres shattered, and the universe thrown open to infinity. The moon was not a crystal orb, nor was it Diana. The moon was another Earth; or, perhaps better to say, the Earth was another moon. There was no distinction between the world above and the world below. Earth and sky were subject to the same laws of nature, laws that embraced change and the lack of it. Within the century following Galileo's lunar observations, a hundred writers would imagine an inhabited moon. And inhabited planets. And other suns with other planets, all inhabited. The Earthly drama was not *The Play*, presided over by the moon, but a play within a play. Perhaps a play within a play within a play within a play.

Not long ago, I read a description by Guy Ottewell of the rising of a crescent moon observed through a telescope. The magnified horizon in Ottewell's field of view was a mountain crest nine miles away. The moon heaved itself into the air, like a "new tarnished-silver mountain budding from the range." In the field of view of the telescope, the trees on the distant horizon were scaled up to the size of the moon. If you sat in the branches of one of those trees, wrote Ottewell (letting his imagination conform to what he saw in the scope), "you would be looking giddily down . . . on the vast desert of the lunar surface, like someone in the dome of St. Peter's staring at its floor . . . you could drop pennies into the craters."

The illusion described by Ottewell was so powerful, the description so compelling, that I resolved to make the same observation. Waiting for moonrise was no problem, but finding an appropriate horizon took some time. It was months before I found myself in a place where the sky was clear to the horizon, and the horizon was sufficiently distant. The time was an hour before midnight, and the moon promised to rise not quite in its last quarter phase. I set up my telescope to point at the place on the horizon where I knew the moon would appear, a range of hills across a dark bay. When at last it came, the moon rose night-side first. The unlit side of the moon was above the crest of the distant ridge for almost a minute before I became aware of its presence. Quite suddenly, there was a scintillation on the black line of the ridge, as if some sort of human activity had begun on the crest of the distant hills, a pagan festival of bonfires and torches. Then the slightly curved terminator came into view, the deeply shadowed dividing line on the face of the moon between moon day and moon night. That line of light gently sagged across my field of view, filling the eyepiece of my instrument with a golden light. It was the rising backdrop of a stage set, a feat of technical theatrics suitable for an opera. Straight up out of the ground came the screen of light—no, not straight up but angling up toward the south, rising and sidling southward. I was immediately in the central highlands of the moon, the mountainous region between the eastern and western "seas." There! . . . a concavity on the line of light and dark, the eastern edge of the Sea of Serenity. And rimming the concavity were the bright sunlit peaks of the Haemus and Caucasus Mountains. The Sea of Vapors disentangled itself from the horizon, and the spectacular range of the lunar Apennines came into view, their great steep eastern flanks catching the full light of the sun; these were the mountains whose Himalayan heights Galileo successfully measured. Next, a line of craters—Plato, Archimedes, Ptolemy, Tycho—the blasted, dusted, wasted surface of a body that has remained unchanged since the violence of its creation. I was out of the shadows now and onto the brilliant sunlit eastern plains. Topography became less sharply delineated. I recognized the huge crater Copernicus, with its prominent central peak, and the young rayed crater Kepler. This was the "wet" side of the moon—the Sea of Rains, the Sea of Moisture, the Sea of Clouds, the Sea of Storms, the Bay of Dew—not really seas and bays at all, but dusty deserts, dark features that give the face to the Man in the Moon. As the full circumference of the moon's bright limb at last broke free of the horizon, I glimpsed the dark blemish of the crater Grimaldi; if you were standing in its bowl it would be lunar noon, the sun high overhead. Was there a crater Galileo on the moon? I couldn't remember.

The whole spectacular show took only a minute. The rising had seemed long and ponderous, but it was over in a minute. The telescope had somehow slowed time as it fattened space. When the rising was over, I took my eye from the scope, and the moon quite suddenly shriveled to become a distant dot of light. Time quickened. I thought of a line from Sir Philip Sidney's sonnet: "With how sad steps, Oh Moon, thou climb'st the skies! / How silently, and with how wan a face." The image of the sonnet did not match what I had just seen. Sidney had missed the telescope by a quarter of a century. The moon did not rise with sad step or wan face. The night I watched moonrise through a telescope, she ascended the air like the Lady of the Lake from dark water, like Titania rising from enchanted sleep, majestic, confident, golden, magical.

But the moon had not risen at all. In fact, the moon's true spatial motion was opposite to what I had seen. It was the Earth that had moved. It was I who was falling with the Earth over the horizon toward the moon, tumbling toward the moon with my hillside and my telescope and my range of distant hills, tumbling toward a new day that already I had seen warming the floor of the crater Grimaldi. The eye deceives. The moon does not rise; the Earth turns beneath the moon. The sun does not travel along the zodiac; the Earth revolves in its huge solar orbit. The universe is not the sleeping Earth's crystalline cocoon, layered about us like a larva's chitin wrap; the Earth, as Galileo guessed, is a winged creature of the night.

❦

It turns out that pieces of the moon have been raining down upon the Earth for eons. Every crater on the moon is a place where moonrock was splattered into space by the impact of an asteroid, and some of that splattered rock fell onto the Earth. The Antarctic ice cap has proved to be an efficient collector of the splattered lunar material. Stones fall from the sky (some from the moon, some from elsewhere) onto the frozen central plateaus of the icebound continent, where they are snowed under and become part of the growing ice cap. The ice flows outward from the central plateaus toward the coast, and the meteoric material embedded in the ice flows with it. In some places the ice reaches the sea and breaks off into bergs, and chunks of Antarctic ice go floating off to warmer waters, and whatever pieces of sky-rock they carry are dropped to the bottom of the sea as the bergs melt. But in other places the flowing ice cap pushes up against coastal mountains. In the dry winds at the back of these ranges, the surface ice evaporates. As the ice evaporates, the meteoric material is left behind on the surface of the glacier, so that half a continent's worth of the rock that had fallen from the sky onto the ice is concentrated in a single place. Millions of years worth of meteorites are gathered by ice and wind for geologists to collect. There, in the ice of the Antarctic mountains, geologists stoop to pick up stones that are identical to rocks the astronauts brought back from the moon—rocks blasted from the lunar surface by asteroids and hurled a quarter of a million miles across empty space into our laps. Not green cheese. Not crystal. Just ordinary rocks. Basalt. Black, piebald, dappled basalt. Chunks, perhaps, of the Lake of Dreams, dropped like pennies onto the Earth.

❦

So Galileo was right. There are no crystalline spheres to hinder the pennies that drop through the sky. This morning in the meadow I rode a planet that orbited with eight others a yellow star. Nine planets skating on the black ice of space, one of them with the moon on her arm. The dusk of moonlight is the mind's natural habitat, said Thoreau. There is a garbled truth in moonlight, a tongue-tied simplicity. In the inarticulate speech of dream there is the Fool's true philosophy. Here is Puck, the moon, with his back to the sun. And here are the planetary players—Snug, Bottom, Flute, Snout, and Starveling—parading in the morning light. *Sweet Moon, I thank thee for thy sunny beams. O grim-look'd night! O night with hue so black! My soul is in the sky: Tongue, lose thy light; Moon, take thy flight.*

17

EARTH,
KIND, MILD

One autumn night many years ago, as I watched through a telescope, I was startled to see the silhouettes of distant geese moving across the moon's bright face. To have observed the birds against the moon in their southerly migration seemed at that time a remarkable coincidence, a personal gift of the night. I later learned that "moon watching" by organized networks of amateur ornithologists was one way the nocturnal migrations of birds were studied in the days before radio-tracking and radar; my gift of lunar geese, apparently, was not so special after all.

Telescopes have a way of "telescoping" distance. The geese I saw silhouetted against the moon were only a few miles away, a hundred thousand times nearer than the moon, but the impression was strong that I had caught the birds far out in space, beating their way across the starlit gulf that separates the Earth from its satellite. And why not? Antoine de Saint-Exupéry's Little Prince took advantage of the migration of a flock of wild birds to make his way from his small planet to the Earth. He was not the only space traveler who harnessed birds for that purpose. Kavi Usan, a monarch of ancient Persia, fastened four eagles to his throne and soared higher than the angels, until the eagles tired and he came back to Earth with a crash. Nimrod, the mighty hunter, after his futile attempt to reach heaven with the Tower of Babel, tried birds—with an equal lack of success. By the seventeenth century, myth was yielding to reason and to natural science, but still, birds—in thought, at least—were harnessed to the task of manned space travel. Savants such as Francis Bacon and John Wilkins seriously discussed the use of birds to assist human flight.

It was Francis Godwin's *Man in the Moone*, published in 1638, that carried flight by fowls to its culmination. In Godwin's book, a Spaniard by the name of Domingo Gonsales travels to the moon and back by attaching himself to a flock of wild swans, or gansas. Godwin was sufficiently versed in the new physics of Kepler and Galileo to tell his readers that at a certain distance into the journey the grav-

itational pull of the Earth ceased and the lines attaching
the traveler to the swans went slack. From that point onward, the birds bore their load as effortlessly as fish in the
water. In space, observed the intrepid Domingo, up and
down and sidelong are all as one. On the twelfth day after
taking leave of Earth, the gansas put Domingo down on a
high lunar hill, where he began to take note of the incredible sights and colors of his new world. Not the least of
the wonders to be seen from that mountain in the moon
was the blue-and-white Earth, suspended like an ornament
in the lunar sky. Domingo confirmed at least one of Copernicus' radical ideas regarding the motions of the Earth,
for with his own eyes he saw the terrestrial globe turning
beneath him.

❦

Photographs of the Earth from space are among the most
beautiful and provocative artifacts of the twentieth century. Once you have seen such a photograph, it is impossible to think of the Earth as other than a celestial object,
a globe caught in orbit by a yellow star, a cloud-dappled
ball tethered on its huge ellipse. Such photographs elicit
the stimulating exercise of imagining the Earth as viewed
from the moon, as Domingo Gonsales saw it. I will take
myself there for the purpose of watching the Earth. I have
a moon map here on my desk; I choose for my vantage
point the 10,000-foot-high central peak of the Crater Plinius, in the strait between the Sea of Serenity and the Sea
of Tranquility, a place comfortably removed by rugged
highlands from the Sea of Crises and the Lake of Death.
Let it be night at the Crater Plinius, in the last dark moments before sunrise.

There is a sense in which sunrise on the moon is
quick. There are no preliminaries, no fanfares, no blush of
dawn to announce the sun. The edge of the solar disk appears abruptly on the horizon, and blades of light go slicing

into shadow. Mountain peaks burn like beacons. Light and dark are delineated with the sharpness of a razor's edge. Every pebble on the slope of my lunar prominence dresses in the harlequin's white and black. But there is another sense in which sunrise on the moon is slow—*little-hand slow*—for one who is used to terrestrial sunrise. The moon turns on its axis under the sun thirty times more slowly than the Earth. A day on the moon is a month of Earth days long. Sunrise on Earth takes two minutes; on the moon, an hour passes before the entire disk of the sun clears the crater rim and it is fully day.

Even as the sun rises, the sky at the Crater Plinius is dominated by the Earth, high overhead—four times bigger than the moon appears from Earth, and four times larger than the apparent size of the sun. The earth is stationary. Its fixed position in the moon's sky is a corollary of the moon's peculiar orbital motion. The moon turns on its axis once a month, in exactly the time it requires to orbit the Earth. This is not a coincidence; tidal forces between the Earth and the moon, acting over eons, have pulled the moon into a synchronous orbit so that it always presents the same face to the Earth. The other side of this synchronous behavior is that the Earth hangs motionless in the lunar sky—there is no "Earthrise" or "Earthset" for an observer on the moon. From my seat at the central peak of the Crater Plinius, the Earth resides permanently near the zenith. At this hour of lunar sunrise, the terrestrial globe is slightly more than half illuminated by the sun. The sunlit hemisphere is daubed with color. I distinguish the massed whites of the Arctic ice cap and the dusty browns of the Asian continent, now moving from day into night. Antarctica is tucked beneath the southern curve of the Earth. Australia is concealed in darkness.

The sun leaves the crater's rim and ascends the sky. For seven days it creeps up from the horizon toward the zenith. The sky is black even though the sun is in the sky.

On the airless moon, the sun and the stars are visible at once; together they climb toward the place where the Earth waits, the guardian of noon. As the sun climbs, the Earth wanes to a thin crescent, closes like a weary eye, presenting at last the full dark circle of its night. At last sun and Earth stand together near the zenith, like a shiny dime and a big English penny displayed on the black cloth of space. The entire floor of the Crater Plinius is illuminated. My mountain peak swims in a bowl of pure light.

Shadows lengthen. The sun and stars leave the Earth perched at the apex of the sky and slip toward the western rim of the Crater Plinius. I wait now for the long hour of lunar sunset. Already the Lake of Dreams, a bay of the Sea of Serenity, reclines in darkness. The Marsh of Sleep, in the highlands between the Sea of Crises and the Sea of Tranquility, has passed into night. The broad basins to the north and south of my promontory fill with shadow. The sun touches the crater's rim and drops behind it. The horizon takes a last bite of the sun and it is instantly night.

Night on the moon is watched over by the great owl eye of the Earth, now slowly widening. I glimpse the Antarctic ice cap in the southern horn of the fattening crescent and watch the continents wake into day. The blue Pacific, with its smudged fingerprint of white cloud, turns to yield its place in daylight to the brown bulk of Asia. Then they come parading out of darkness—Australia, China, India, Arabia, Africa, and Europe, the fat blue S of the Atlantic, the prow of South America steaming into day with North America in tow—whole nations awakening to daylight again and again as the Crater Plinius passes its lunar night.

The Earth wanes from full. The owl's eye slowly closes. Night appears on the planet's western limb, and again I watch continents glide from daylight into darkness. When twenty-nine-and-a-half Earth days have passed, the sun's disk reappears on the crater's eastern rim. A new day begins. The Earth, perched at the top of the sky like an owl on the roof beam of the world, has blinked once.

❦

There are species of higher animals that live out their lives in permanent night. Sea cucumbers, rattails, angler fish, brittle stars, and gulper eels inhabit the ocean abyss at depths greater than 4000 feet—beyond reach of sunlight— feeding upon organic matter that drifts down from the realm of day and night, food derived ultimately from the energy of a star, falling out of the darkness like a gentle rain. Insects, spiders, crayfish, and salamanders, ghostly white and eyeless, live in the unmitigated darkness of deep caves, relying for their meals upon the droppings of occasional visitors—bats, birds, and mice—that come and go from daylight. Moles, mole crickets, earthworms, and centipedes forage for organic scraps in the murky microcaverns of the soil. All of these creatures, though adapted to darkness, are part of food chains that are anchored at the top by photosynthesizing plants. There is one exception to this rule. At recently discovered hydrothermal vents along rift valleys of the ocean floor, there are communities of animals that live *totally* independent of sunlight. Hot, mineral-rich water, heated by contact with the Earth's molten upper mantle, bubbles up from fissures in the stretched crust. Minerals precipitating from the hot water build pagoda-like structures of rock. About these smoky chimneys miles beneath the sea, bacteria, tube worms, giant clams, and white crabs extract a living directly from the energy of the Earth's core in a night absolute and pure.

The creatures who eschew light are throwbacks in the story of life. They are dropouts and atavists, backwards evolvers that still carry the badges of a sighted past. Moles have rudimentary eyes behind the opaque flaps of skin that shut out soil. Cave-dwelling crayfish have lost all trace of eyes but retain eyestalks. Cave salamanders have pigment spots on the sides of their heads where once their ancestors had eyes. Atavists and throwbacks. Retrograde experiments in evolution. Sight has always been the future. The Earth is a planet with a rage to see and be seen. The eyes in my head are the *pièce de résistance* of evolution, the planet's eon-tempered instrument of self-inspection. Encoded in neural pathways of my brain is an image of a blue-white globe suspended in the night, revolving in its conical shadow, waxing and waning like the moon, a sphere of refractory dust gathered on a core of iron, damp with sea water, misted with volatiles, wrapped with a film of light-sensitive and light-emitting organic matter.

Viewed from the moon, the night side of the Earth is faintly lit by reflected moonlight. But the Earth also shines by its own pale emanations: glowworms and fireflies, luminescent toadstools and flashlight fish, light-emitting plankton and bacteria, plants and animals flashing and glimmering, radiating a cold biological light, proteins combining with oxygen to produce a light more refined than the light of any star. Certain fungi of the Far East can be seen from far off by their own shining. The lips of the "megamouth" shark are lined with hundreds of tiny lights, twinkling like a fairground's string of bulbs, enticing plankton into the gaping mouth. Some starfish, if threatened, will shed a glowing arm to distract the attacker. Male fireflies of the species *Pteroptyx malaccae* of Southeast Asia gather in special trees each evening and begin to flash their feeble lights, at first randomly, then increasingly in synchronization, until at last the entire tree blazes with insects pulsing in unison, making a beacon bright enough to attract females of the species from acres around. If all of the Earth's light-makers, microscopic and macroscopic, of sea and land, could synchronize their lights like *Pteroptyx malaccae*, blinking together, who can say if, seen from a prospect on the moon, the Earth would not glow like a will-o'-the-wisp, self-announcing, asking to be seen, a pale biological star in its own right, a jack-o'-lantern planet burning in the night.

❧

"I consider it a sign of human weakness," wrote the Roman natural historian Caius Plinius Secundus, "to inquire into the figure and the form of God. For whatever God be, and wherever he exists, he is all sense, all sight, all hearing, all life, all mind, and all within himself." Pliny was careful not to circumscribe God's figure or form. Everything that exists was the subject of his study. Not one of the senses was neglected. The sights, sounds, tastes, smells, and tactile sensations of this world were each a partial and equally precious revelation of the Earth's divinity. Of these divine trifles Pliny constructed a natural history of thirty-seven volumes. The first, the preface, is dedicated to the Emperor Titus Vespasian; the second is on the figure of the world, the elements, and the heavenly bodies; the third, fourth, fifth, and sixth volumes are on geography, and the seventh concerns itself with the inventions of man. The remaining books treat mammals, fish, birds, insects, plants, medicines, minerals, and gems. No sand-encompassed oasis, no fish that swam in a faraway sea, no cave, no spring, no pebble or rock, was beyond Pliny's interest. The Earth turns before us in Pliny's *Natural History* as if we were watching from the moon, exposing night and day, seas and lands, deserts and tundra, with earth, water, air, and fire arrayed before us in an ascending stair, reaching up to the sphere of the moon and tipping the central peak of the Crater Plinius with a divine flame, a votive offering to tranquility and serenity.

If Pliny had stood on the lunar mountain that bears his name and seen the Earth hanging at the zenith of the sky like a lamp hung at the apex of the Pantheon's dark dome—hanging there throughout the long lunar day, turning thirty times in its conical cap of night, godlike and majestic—that wonderful sight would only have confirmed for him the divinity of the world. Pliny's song to the Earth is one of the most beautiful passages in his *Natural History*. "She receives us at our birth," he writes, "nourishes us when we are born, and ever afterward supports us; lastly, embracing us in her bosom when we are rejected by the rest of nature, she then covers us with a special tenderness." Earth, water, air, and fire, glowing in the night; God's eye at the apex of the temple, watchful and consoling. Pliny continues: "The water passes into showers, is concreted into hail, swells into rivers, is precipitated into torrents; the air is condensed into clouds, rages in squalls; but the earth, kind, mild, and indulgent as she is, always ministers to the wants of mortals."

18

WAITING
FOR THE COMET

P raise *this* world to the Angel, says the poet Rainer Maria Rilke in the Ninth Duino Elegy. Do not tell him the untellable. Tell him *things.*

And so be it. The time has passed when we could hope to speak with certainty of the Alpha and the Omega. Space and time extend beyond the tips of our tongues. The galaxies go tumbling off the precipice of speech. Perhaps, says Rilke, we are here only to say, house, bridge, fountain, gate. It is enough. Look, says the Angel, here is a house. "House," I repeat, praising. And a bridge. "Bridge," I say. And a fountain. "Fountain." And a gate. "Gate."

Here, says the Angel, is a comet.

❦

In 1948 Halley's Comet turned the long dark corner of its ellipse far out beyond the planet Neptune. Unseen and recorded only by the blind eye of the astronomer's calculations, it leaned into its sunward curve (in the phrase of the poet Ted Hughes) "like a skater on the thin ice of space." In October of 1982 the returning comet was recovered telescopically by a Cal-Tech team of astronomers using something called the "Space Telescope Wide-Field Planetary Camera Investigation Definition Team Charge-Coupled Device" mounted at the prime focus of the 200-inch telescope on Mount Palomar. At that time, Halley's Comet was a round billion miles from Earth, plummeting pell-mell toward the sun. As I write, the comet is racing across the orbit of Saturn, still invisible to the eye, a ghostly presence in the dark space between Canis Minor and Orion. In the winter of 1985–86 it will lash its white tail in the evening sky, dash around the sun, and begin its heavy climb back up into the solar system's dusky attic.

Halley's Comet returns every seventy-six years. It will not be as bright this time around as last. In 1910 it passed very close to the Earth and presented a full profile of its glowing head and sweeping tail. However, in 1986 the Earth will be on the wrong side of its orbit, and we will not get a broadside view of the comet. The imminent appearance of Halley's Comet, then, does not promise to be the spectacular show it has sometimes been in the past, the perhaps not even as bright as some other comets of recent years. But *any* visitation of Halley's Comet must excite interest. It is the only bright comet with a period of less than a century. (Comet Kohoutek, for example, won't be back for 80,000 years.) It is the comet whose reappearance was predicted by Newton's friend Edmund Halley. Relying upon Newton's theory of gravitation, Halley guessed that the bright comet of 1682 would reappear near the end of 1758. The astronomer did not live to see his prediction—and Newton's theory—dramatically confirmed when, on Christmas night 1758, the comet was rediscovered with a small telescope by a German farmer named Georg Palitzsch. We now know that Halley's Comet is the same comet that is stitched into the Bayeux Tapestry of 1066 as a sign of God's favor for William's conquest. It is the same comet that frightened Europe when it appeared in 1456, shortly after the fall of Constantinople to the Turks. In 218, it was the "fearful flaming star" that preceded the death of the Emperor Macrinus. Attempts have been made to link the comet's apparition of 12 B.C. to the Star of Bethlehem.

Before Newton's theory and Halley's successful prediction, it was the *unexpected* aspect of a comet that made it a portent and a thing of wonder. Today, it is precisely the *predictable* character of the comet that makes it such a curiosity. Mark Twain was born during the visitation of 1835. He said he came in with the comet and would go out with it, and he died, as he wished, as Halley's Comet made its 1910 dash around the sun. It is a good story, this story of Twain's birth and death. A comet's apparition is, after all, a good time to be born and a good time to die, and seventy-six years is a respectable span of life. My father was born during the visitation that coincided with Mark Twain's death. He was not able to match the span of his life to the period of the comet, but I will be there in his place in 1986, when Halley's Comet drops out of the night and falls toward the sun, chaos bound to an elliptical track, diving with the roller coaster's lurch and scream.

❦

First, there was Kohoutek. In March of 1973, Lubos Kohoutek, working at the Hamburg Observatory in Germany, discovered two faint comets on a photographic plate. One of these comets was on an orbit that would take it very close to the sun. It seemed inevitable that it would be very bright, perhaps as bright as the planet Venus. This was the first time such a potentially luminous comet had been discovered so long before it reached perihelion (the point where the comet comes closest to the sun). The time of anticipation would be long and especially keen.

Kohoutek became a media event. The papers ballyhooed its coming. Paperbacks appeared on the newsstands. Cranks in public places passed out leaflets announcing the end of the world. The Franklin Mint produced "once in a lifetime" keepsake plaques in pure silver. The QE2, the greatest liner afloat, sold tickets for a "comet cruise," complete with lecturing scientists and telescopes. Everyone, it seemed, wanted in on the act. Nineteen-seventy-three was, after all, the year of Watergate and Spiro Agnew, a nadir of our national spirit. We would make of Kohoutek an American comet, and it would appear like a skyful of tinsel just at Christmas. It would be a star-spangled visitation, its head and tail a dazzling exclamation mark to the State of the Union. This would not be one of those faint half-a-dozen-every-year comets of the astronomers. This would be an ICBM of celestial fireworks, an American affirmation of uncounted megatons, our prime-time Christmas Special, our Super Bowl.

With sublime discretion, Comet Kohoutek made its appointed rounds. Inexplicably, the comet failed to develop its expected brightness. It was perhaps a virgin comet, approaching a star for the first time, its early brightness blown off by the solar wind. During exquisite deep-blue dawns of mid-December, the comet played hide-and-seek. I pursued it with binoculars but got no more than a glimpse. In early January, Kohoutek slipped quietly around the sun and reappeared in the pale pink sky of evening, a faint blur in my binoculars, a perfect twin to the galaxy in Andromeda. As the days passed it climbed above the pines on the western horizon to creep past Venus and Jupiter just at the time of a beautiful conjunction of those two planets. Toward the end of the month it glided on to a rendezvous with the slenderest of crescent moons.

Kohoutek's visitation was a thing of hints and traits. The paperback publishers and cruise-ship entrepreneurs were disappointed. But Kohoutek was a perfect comet and, like all perfect and beautiful things, was known not as we wished it to be but as it is.

❦

". . . And when I said, 'The limits of my language are the limits of my world,' you laughed. We spoke all night in tongues, in fingertips, in teeth." The line is from a poem by Robert Hass. Tongues, fingertips, and teeth! This is the language of comets, wordless, night-spoken, dawn- and dusk-bracketed, a language with a syntax of pale light and a grammar that arches voids, slides along ellipses, gathers force, and blows away in the solar wind, a lover's language of nod and nuance that stretches the circumference of our world.

March 1976 did not come in like a lamb. Rather, it arrived in leonine fury, with snow, ice, and sleet. And somewhere out there beyond the lion's mane of heavy clouds was Comet West. The comet was discovered in late 1975 by Richard West, on a photographic plate made in Chile when the comet was still invisible to the naked eye.

It was clear from the beginning that Comet West would be brighter than Kohoutek, perhaps one of the brightest comets of the century. But there was no publicity. The entrepreneurs, once bitten, were twice shy. By the third of March the comet had passed the sun and should have been visible to naked-eye observers in the northern hemisphere. But for a week the sky was overcast. As the days passed, there was a distinct possibility that we would miss the comet altogether. Behind the clouds, Comet West continued on its parabola, defiantly shaking its tail like a gazelle that knows its speed will let it outrun the lion.

During the daylight hours of March 5, the clouds melted away. The sun burned briefly against the snow. I looked up and knew Comet West was there, lost in sunlight, its faint light no match for that of our furious star. Then, again, the clouds joined. I waited, one, two, three more days for a clear dawn. On the eighth of March at 5 A.M. I was in the back yard with binoculars. There were only a few scattered wisps of cloud in the east. I scanned the horizon. No comet. Reluctantly, I decided that Comet West was too faint to be seen with binoculars, perhaps even fainter than Kohoutek. Then, as I was about to give up the search, a tiny cloud drifted to the south and there was West, as bright as a first-magnitude star, easily accessible to the naked eye. It looked exactly as a comet should, like a textbook photograph. The coma burned in reflected sunlight. The long tail stretched straight up from the horizon for three or four degrees. The comet was much brighter than I expected it would be, an object of exceptional beauty. I stood watching in the cold dooryard until the pinks and blues of dawn drowned the comet's light.

"The palm at the end of the mind," begins a poem of Wallace Stevens, "beyond the last thought, rises in the bronze distance." Comet West was the palm at the end of the mind, the intuition toward which the intellect spontaneously turns. Comet West hung elusively in the morning sky like a fleeting thought, like a liquid image dissolving in the darkness at the edge of a dream. It was the gold-feathered bird of Stevens's poem, singing in the palm at the edge of space, its feathers shining, its fire-fangled feathers dangling down.

The evening of the twelfth of March, 1976, was clear and cold. I knew when I went to bed that the next day would dawn with a horizon as black as slate. I also knew that this would possibly be my last chance to see Comet West, now racing from the sun and chasing its own long sunblown tail. Still, when the alarm sounded I turned it off and fell back to sleep. I would regret my laziness. My students told me later that the comet was brighter than ever that morning, the tail longer, stretched against the stars by the exceptional clarity of the dawn, fire-fangled, shining.

❦

Comet West has a period of 15,000 years. When last it was here our Cro-Magnon ancestors were cowering in caves to escape the rigors of the Ice Age. Comet West's orbit takes it forty times farther from the sun than the planet Pluto. From the languid repose of its apogee, the sun appears no more prominent than any other star. The comet turns that distant pole unseen, lit only by starlight.

Comets have been called "dirty snowballs," a rough metaphor for masses of frozen water, carbon dioxide, ammonia, methane, and dust. This is the same stuff that life is made of. Take a comet, heat it up, put a spark to it, and it will puff and heave like an anthill; it will crawl with life like a rotten melon. The stuff of comets is the stuff of our own bodies. And a comet's brief and predictable appearance in the Earth's sky is the stuff of our imaginations. Language bends and stretches to encompass it.

Comets have their origin in a cloud of light, volatile materials that surround and enclose the solar system—the so-called Oort cloud, a spherical aura of life-stuff blown off by the awakening sun. In the outer reaches of the cloud there are billions of comets waiting to fall toward the sun. Once a comet begins periodic visits to the inner solar system it must eventually be blown away, dispersed, evaporated. A comet's tail is a comet diminished. It is matter pushed off by sunlight. With every apparition, Halley's Comet withers. But a billion other comets wait in the wings to trip the light fantastic. "There is a higher mystery," says D. H. Lawrence, "that doesn't let even the crocus be blown out." A higher mystery? Or is it a lower mystery? A mystery in language, in the soil, in the seed. A mystery that even now is moving invisibly in the dark space between Canis Minor and Orion. Soon it will transit the Pleiades. We wait for Halley's Comet, for the big one. This is the house, this is the bridge, this is the fountain and the gate. Praise the world to the Angel; tell him *things*. Tongues, fingertips, and teeth.

19

HOW SLOWLY
DARK

L ate last night I walked to the plank bridge over the Queset Brook near my home in Massachusetts. The sky was dark and clear, far darker and clearer than we have any right to expect here so close to a large city. In the black water that flowed beneath the bridge I saw a glint of light that seemed to come from deep within the stream, like the spark in a gypsy's crystal ball. For a moment I searched my mind for a possible source of the light—a reflecting grain of quartz, perhaps, or a tiny luminescent plant or animal. Then, suddenly, I knew what it was that I was seeing. It was Capella, the brilliant star now at the zenith, reflected in the dark water of the brook. So! the stars flow too. The brook was light-years deep. Here beneath the bridge was another universe, flowing in the dark water. Galaxies whirling in the stream like the egg cases of caddis bugs. Nebulas of stars keeping company with dragonfly nymphs. Mosquito larvae feeding on the dust of novas.

So bright a night! No moon, not even planets to mask the sky's faintest lights. The stars of Andromeda were still high in the west; I looked among them for the Great Galaxy—M31 in Messier's catalogue of fuzzy spots—an object I have seen with the naked eye only on the best of nights. And I found it, quickly, north of the star Mirach and south of Shedir in Cassiopeia, the great Andromeda spiral, the companion galaxy of the Milky Way, drifting beyond the stars. What I saw was not a source of light, really; it was more like a raindrop on a pane of glass, shaping the darkness rather than diminishing it. It occurred to me that this would be an excellent night to go looking for galaxies with the telescope. M31 in Andromeda was readily accessible even to the naked eye. M33, the Pinwheel Galaxy, and the elusive M74 were nearby. The constellations Ursa Major and Leo were rising in the east; in a few hours they would be well placed for observation with their clutch of spirals. I walked to the college and cranked open the aperture of the observatory dome.

Ye littles, lie more close! A line from a poem of Theodore Roethke, a poem I have quoted more than once in these pages. That line seems to me an appropriate prayer for a searcher of the night. The night speaks a language of simple syllables, the sort of language Bernard longs for in Virginia Woolf's *The Waves*, "a little language such as lovers use, broken words, inarticulate words, like the shuffling of feet on pavement." I have often shown the Great Galaxy in Andromeda to friends and students. They have sometimes expressed a kind of disappointed surprise at the misty smudge of light they see in the eyepiece of the telescope. Perhaps they rightly expected something more from an object called the "Great Galaxy," or perhaps they have seen the magnificent long-exposure photographs of this object prepared by major observatories and are disappointed that the galaxy "in real life" fails to live up to its 8 × 10 glossies. Observing the deep sky with a telescope means listening for the muted sounds of distant eloquence, sounds muffled by light-years, broken, inarticulate sounds, dream sounds, whisperings, sweet nothings. *Lie close, ye littles!* I swung the 14-inch Celestron telescope toward Andromeda.

The galaxy in Andromeda is the nearest and brightest of all the spiral galaxies. It lies 2 million light-years beyond the Milky Way. Last night, in the telescope, a pale oval of light reaching two or three degrees across the sky surrounded the bright central nucleus. Seldom have I seen Andromeda's galaxy more spectacularly displayed. M33, the Pinwheel spiral in Triangulum, was more difficult. I scanned the tiny constellation with a wide-field, low-power eyepiece, looking for the slight modulation of darkness that would announce the galaxy. It took a considerable time to find what I was looking for, but once found, the object was easily retained. There are perhaps 10 billion suns in the Pinwheel Galaxy, flowing through the night like stars reflected in a dark stream. M74 in Pisces is another face-on spiral; observatory photographs show two tightly coiled arms of brilliant stars. I knew from previous experience that this particular object would be a challenge to find. I began with the telescope on the star Eta Piscium and then edged into the abyss of darkness that separates this star from Sheratan in Aries. I believe that I saw M74, a whisper of light so transparent that only the memory of a photograph was its validation.

Now I rolled the dome of the observatory to face the stars of the Big Dipper, rising in the northeast. The two brightest galaxies in the Dipper, M81 and M82, were easy to find, as always—paired ellipses of white light, one fat, one thin. The flat-on pinwheel spiral, M101, at the end of the Dipper's handle, took more time; it was a barely perceptible diminution of the darkness. But the Whirlpool Galaxy, in nearby Canes Venatici, was the best I had ever remembered seeing it, distinctly defined in the field of the telescope, with a hint of spiral arms, and with the conspicuous companion nebulosity dangling from its tail.

There were other galaxies that I saw last night, other distant and indistinct smudges of light, other spirals of stars more imagined than seen, other hints of a universe that extends beyond my speaking and my knowing. And all of the galaxies, all but the very nearest, were racing away from me, racing into darkness, stretching and dimming their light, blackening night, pulling the universe thin like taffy. The universe is expanding! The Whirlpool Galaxy in Canes Venatici is 35 million light-years away and receding from the Milky Way at 340 miles per second. Tonight that galaxy will be 30 million miles farther away than it was last night, a tiny inflation on the scale of the galaxies, but patient, inexorable, and ultimately exhausting.

☙

There is now general agreement among astronomers that the past and future evolution of the universe was determined in the first few moments of the Big Bang. *Yea,* says the *Rubáiyát* of Omar Khayyám, *the first Morning of Creation wrote what the Last Dawn of Reckoning shall read.* Fifteen billion years ago a momentous explosion flung into being all that exists today—space, time, matter, galaxies, and stars—and from that moment the fate of the universe was sealed. Astronomers are not certain, however, what that fate will be. One possibility is that the universe will continue forever its present expansion, infinitely diluting the density of matter and energy, augmenting darkness, snuffing out at last every faint light. The other possibility is that gravity will ultimately slow and stop the present inflation and pull the universe back together, back to the infinitely dense fireball of Creation, like dry bones, dry bones coming together in noise and commotion, each one to its joint with the flesh upon them, *hear Ye the word of the Lord.*

In principle, it should be easy to decide whether the universe will end in eternal night or singular noon, in the infinitely thin dispersal of its substance or in the re-creation of the flash of the Big Bang. If the present average density of the universe is greater than three protons for every 1000 liters of space, then gravity will eventually prevail and the universe will collapse back toward its beginning. On the other hand, if the average density of the present universe is less than this critical value, the expansion will continue forever. Current estimates for the average mass density fall well below the critical density. But those estimates are based on an accounting of *luminous* matter, a toting up of the masses of the visible stars and the bright nebulas in visible galaxies. If *nonluminous* matter accounts for a significant portion of the mass of the universe, then estimates of the average density could be wide of the mark. Astronomers are beginning to suspect that a good part of the mass of the universe might, in fact, be subtly hidden from us, in black holes, in dead stars, in objects of substellar mass, in massy neutrinos, perhaps in yet un-imagined forms of dark matter. The issue is unresolved. Take your pick. Death by fire or death by ice? Light or dark? Bang or whimper?

Consider the second option first. At the moment, the death of the universe by the dispersal of its substance seems the more likely of the two alternatives. The galaxies are racing away from one another, and the gravitational elastic that holds them together is probably too weak to stop their long slide into darkness. In 20 billion years the Whirlpool Galaxy will have doubled its distance from the Milky Way. But there will be no one here on Earth to go searching for that fainter light on a starlit winter night. By then, the sun will have exhausted its energy and died, briefly swelling in its death throes to the status of a red giant. The Earth will have been seared and left to freeze in darkness, a lifeless rock in orbit around an extinguished star. Within some tens of trillions of years, all of the stars in the galaxies will have used up their energy resources and expired. Night's lights will have ceased to shine. Galaxies of dark starlike masses will whirl unseen—cold, ghostly pinwheels, black on black. Thousands of trillions of years will pass, and dead stars, moving at random in the dead galaxies, will encounter one another and gravitation-ally kick their planets free. Occasionally, if the encounter is close enough, one of the colliding dead stars will itself be flung from its galaxy, so that slowly the galaxy will lose its mass. Galactic evaporation! After as much as nine-tenths of the mass of a galaxy has evaporated, gravity will draw whatever remains into a great central black hole. After a billion billion more years pass, the universe of spi-ral galaxies will have been replaced by a universe of super-massive black holes and unattached nonluminous stars adrift in limitless space. After a trillion billion billion more years, the protons that make up the matter of the dead stars will begin to decay, and the universe of stray nonluminous stars will distintegrate to become a rarefied gas of electrons and positrons; stars spontaneously disassembling and dispersing their substance. At last the great black holes, which will be all that remain of the galaxies, will lose their matter by quantum evaporation. After a number of years equal to one followed by one hundred zeros, the universe will be an extremely diffuse gas of electrons, positrons, neutrinos, and photons, drifting ever farther apart, the candle of Creation snuffed, the wick crushed between the fingers, the soot blown to the wind.

In the foregoing scenario, the universe ends in cold and dark, its density approaching zero, its substance reduced to a ghostly stuff infinitely dispersed. But what if there is sufficient hidden matter in the universe to give gravity the ultimate snap that will pull things back together? Then for a hundred billion years the universe will continue to expand, but ever more slowly, until at last the galaxies are poised at the limit of their flight, as a ball thrown into the air pauses instantaneously at the top of its trajectory. As the galaxies start their fall back together, they will consist of dead stars and supermassive black holes. The universe will contract, slowly at first, but accelerating. Fifteen billion years before the ultimate crunch, the universe will have contracted to a volume at which its energy density is the same as it is today. Photons of light will gain energy from their fall, and the universe will heat up. A million years before the crunch, energetic photons will disassociate interstellar hydrogen into protons and electrons. A year before the crunch, stars will break up and explode. At about the same time, the supermassive black holes that were the collapsed cores of the galaxies will begin to collide and coalesce. They will annex the diffuse matter that is dispersed about them. The black holes will finally combine into a single black hole that is coextensive with the universe itself, a black hole that continues to contract toward a state of infinite density and zero size, a universe reduced to the size of the head of a pin, to the size of the point of a pin, to the size of a single atom, a universe shrinking without limit, stars and galaxies washed down the drain and the drain pulled down after them.

Now the equations of physics break down, and the crystal ball of theory goes dark. Perhaps that is the end of the story. Perhaps the universe will bounce back from its state of infinite density and begin another expansion, another Big Bang, repeating Creation, with new stars, new galaxies, new starry nights. Or perhaps our universe is only one bubble in a froth of universes, expanding and popping, puffing up and collapsing, one bubble in an infinite matrix of bubbles that boils from the texture of space-time like fizz from champagne.

❦

I stand by a low fire / Counting the wisps of flame, and I watch how / Light shifts upon the wall. I bid stillness be still. / I see in evening air, / How slowly dark comes down on what we do. Lines from Roethke, wisps of flame. Stars reflected in streams, galaxies that are smudges of light in the eyepiece of a telescope, shadows on a wall. A language of little sounds, like fcct shuffling on pavement, whispered revelations. Some weeks ago I was walking in the woods near Morse Pond. Quite suddenly I was surrounded by popping sounds, Rice Krispie sorts of sounds, like sleet falling onto the trees out of a clear sky. It was the witch hazel, that most remarkable tree, so out of step with the season, blooming in November, tiny golden flowers exploding like popcorn along the gray branches to scent and color the dismal month. The deciduous woodlands had closed down for the season, but here still they rattled with life. Life exploding like stars, yellow flowers spiraling like galaxies, matter animated, irrepressible, exuberant.

Those smudges of light I searched for last night in the dark spaces of Andromeda and Ursa Major, those Milky

Ways reflected in the pool of night, are not galaxies racing into oblivion like stones tossed. Surely they snap, crackle, and pop with life. Everywhere, matter burns with life's flame. Stars shimmer in their green auras. Planets puff with spores. Galaxies respire. The Whirlpool Galaxy and the Pinwheel Galaxy are prayer wheels spinning out supplications. The Great Galaxy in Andromeda sails like an ark into the great good night, with every creature two by two. The fate of the universe is the fate of life.

The Lord told Ezekiel to shave the hairs from his head. A third part he was to burn in the fire. A third part he was to cut into pieces. A third part he was to scatter to the wind. But a small number of the hairs he was to bind into the skirt of his cloak, to be the surety of Israel. And so, a third of the galaxies will be consumed by fire, and a third will be cut into pieces, and a third will be scattered to the wind. The universe will end in fire or in darkness. Stars will evaporate like pools of rainwater on the flanks of volcanoes; galaxies will collapse into ponderous knots. And we are the hairs sewn into the hem of the cloak.

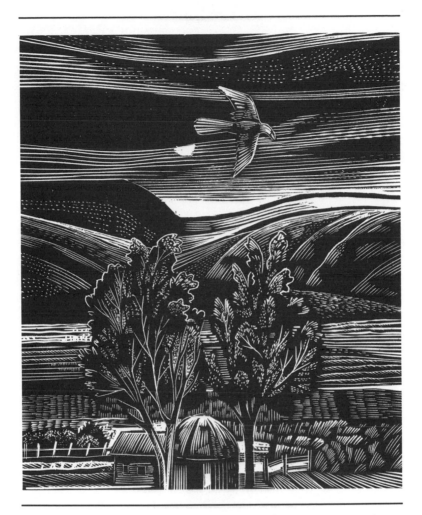

20

THE BIRD AND
THE FISH

T he migrant bird longs for the old wood: The fish in the tank thinks of its native pool." With these words, the fourth-century poet T'ao Ch'ien left the city and retired to a rural place, a patch of the Southern Moor, a thatched cottage, elms and willows by the eaves, peach trees and plum trees in the garden beside the three paths, where on summer evenings he might watch the Milky Way rise above the eastern hedge. Beneath the hedge, in the cool light of 500 billion stars, chrysanthemums spread pale blossoms.

T'ao Ch'ien was born into a family of modest influence at the time of the Tsin Dynasty. His great-grandfather was the Duke of Ch'angsha; his father and grandfathers were officials in the bureaucracy of the empire. T'ao himself embarked upon a career in the civil service but soon tired of the facades and trappings of office. He refused to put on the official belt when reporting to his superiors. He declined to kowtow for five pecks of rice a day. At the age of thirty-three he turned his back on a life of quiet desperation and sought his Walden Pond. Like Thoreau, he built his house near a place of human habitation, but he did not hear the sound of horse or coach. "A heart that is distant," he wrote, "creates a wilderness around it."

At the latitude of T'ao's retreat on the slope of Mount Lu south of the Yangtze River, the eastern horizon at midsummer lies parallel to the plane of the Milky Way Galaxy. As the Earth turns toward the night, the Milky Way rises on the distant skyline like a gathering of cloud. On summer evenings T'ao watched the fullest stream of the central Galaxy lift above his hedge, from Cygnus in the north to Scorpio in the south, or, as T'ao would have expressed it, from the Celestial City of Tien Tsin to the Azure Dragon. His chrysanthemums bloomed in the milky light.

The 10,000 miles that separate the slope of Mount Lu from the shore of Walden Pond are a single step along the white path that bridges heaven and Earth. Fifteen hundred years after T'ao, Thoreau watched the Milky Way rise on summer nights, just as T'ao saw it rise—the same bold stars, the same patterns of the constellations. The poet of the Southern Moor and the naturalist of Concord were neighboring citizens of their Galaxy: two distant hearts, one wilderness. The Earth was their hermitage, surrounded by a wilderness of stars.

❦

"I pluck chrysanthemums under the eastern hedge," wrote T'ao Ch'ien, "then gaze long at the distant summer hills. The mountain air is fresh at the dusk of day: The flying birds two by two return." Tonight, just at dusk, I watched the two great birds of summer return, the swan and the eagle—Cygnus and Aquila—skimming the eastern air. This is the part of the night sky I know best, the sash of the summer Milky Way flowing in two bright streamers from north to south and gashed by the Dark Rift. On warm summer nights I have explored the Milky Way with binoculars and telescope. I have pored over observatory photographs of the extraordinary objects that float in this river of light, and I have sought them in the sky. The summer Milky Way holds delicate gifts. Here is the blue and gold binary star Albireo, a single point of white light to the naked eye, but two stupendous suns in the small telescope, one star 760 times more luminous than our sun, the other 120 times brighter, sapphire and topaz; fifty-five solar systems could be lined up in the space that separates this double gem. Here is the star known as 61 Cygni, which I have often seen with the naked eye on nights when I have been far from city lights, a nondescript star, one among billions in the body of the Swan, famous as the first star whose distance was measured by direct triangulation, by Bessel in 1838; 61 Cygni is eleven light-years away, 65 trillion miles, 10 billion times the distance from T'ao's retreat on Mount Lu to Walden Pond; of the naked-eye stars, it is the fourth closest to the Earth, after Alpha Centauri, Sirius, and Epsilon Eridani. Here is the Veil Nebula, a lace of filamentary gas in the shape of a question mark, as delicate as the mists that cling to the slopes of Mount Lu; once, on a dark hillside in the west of Ireland, I glimpsed the brightest part of the Veil Nebula in powerful binoculars; it is the shredded envelope of an exploded star; 30,000 years ago a star 1500 light-years away blew itself to bits. Here is the Dumbbell Nebula of Vulpecula, an easy object in my 8-inch telescope, a puff of stardust, a smudged fingerprint on the globe of the sky. And here are the two delicate constellations Sagitta and Delphinus, the dart and the dolphin; I regard them more fondly than the constellations of brilliant stars that bracket them on either side; Sagitta and Delphinus are mere hints of constellations, four and five fifth-magnitude stars respectively, constellations for the connoisseur of night's faintest lights. Tonight I collected these gifts of the summer Milky Way with the naked eye and with binoculars. The region is banked with stars, puddled with light, shoaled with stardust. There are fish here that flick the light-years with their tails, and birds with galaxies tucked like pinfeathers under their wings. I longed for the old wood; I wished for the native pool. I stood in the dooryard and the wilderness was mine.

❦

T'ao Ch'ien was not an anchorite. Nor was he an ascetic. He wore no hairshirt. He did not go to the Southern Moor to escape the world, but to find it. His pleasures were in nature, poetry, books, wine, and family life. He gardened. He took pride in the chrysanthemums that grew beneath the eastern hedge and in the lettuce that was still moist at midsummer. After the ploughing and the sowing, there was time to sit and read his books, and time to play the lute. A gentle rain blew up from the east, accompanied by a "sweet wind." T'ao's eyes wandered over the pictures in his book, renderings of hills and seas and distant skylines and forests stilled by fog. "At a single glance," he wrote,

"I survey the whole Universe. He will never be happy, who such pleasures fail to please!"

Tonight in a sweet wind I stood beside my door and swept Sagittarius with binoculars. I was looking into the very nucleus of the Galaxy, where, if astronomers are right, a black hole hides in banks of stars. Once I swam in a dark cove in Maine and with every stroke stirred up planktonic luminescence—millions of dinoflagellates, one-celled plants that flash microscopic strobes when disturbed—the sea shimmered in eddies of light; Sagittarius through binoculars was like that. I stopped my sweep of the constellation just north of the star Gamma Sagittarii. It was one of those perfect nights when even without optical aid the star clouds of Sagittarius are prominent. The field of my instrument was monstrously white with stars. What I saw was not a mist rising on the garden hedge; it was a galaxy of worlds without end, a labyrinth with a minotaur at its center, a labyrinth with 500 billion blind turns, and no one has the thread.

No wonder our ancestors imagined they saw birds and fish in the night sky. The swan and the dolphin. The eagle and the whale. The little horse. The water carrier. The lyre. The wolf. There are eighty-eight official constellations—eighty-eight comfortable projections of ourselves onto the wilderness of night. Some of the constellations were ancient when Moses heard the voice in the burning bush. Forty-eight were listed in the *Syntaxis* of Claudius Ptolemy. The oldest extant star map was compiled by the Chinese astronomers Shih Shen and Kan Te in the fourth century B.C. The Chinese did not study the stars with the disinterested curiosity of the modern astronomer. According to Robert Burnham, the Chinese studied the natural order of the night to stabilize the social order that was based upon it, or to accommodate one's personal life more closely to it. Well then, how shall I accommodate *my* life to the star-powdered clouds of Sagittarius? There is a problem of scale, isn't there? Do I accommodate myself to the Galaxy the way the microbe accommodates itself to the gullet of the whale? There was a time when we could believe that the Milky Way was a bridge between heaven and Earth. No more! In the Milky Way there are a thousand billion Earths and one monstrously indifferent heaven.

Yet, I am a child of the Milky Way. The night is my mother. I am made of the dust of stars. Every atom in my body was forged in a star. When the universe exploded into being, already the bird longed for the wood and the fish for the pool. When the first galaxies fell into luminous clumps, already matter was struggling toward consciousness. The star clouds in Sagittarius are a burning bush. If there is a voice in Sagittarius, I'd be a fool not to listen. If God's voice in the night is a scrawny cry, then I'll prick up my ears. If night's faint lights fail to knock me off my feet, then I'll sit back on a dark hillside and wait and watch. A hint here and a trait there. Listening and watching. Waiting, always waiting, for the tingle in the spine.

❦

In the month of June the grass grows high, And round my cottage thick-leaved branches sway. There is not a bird but delights in the place where it rests: And I too—love my thatched cottage. I put aside the poems of T'ao Ch'ien and listen to the sounds of the night. The crickets and the owl. The cicadas and the east wind. I recall a night in the spring, a night of full moonlight, when I heard the male woodcock in his ascending fervor, the strange trilling of the wings, like a resonance, and the clear bell-like warbling at the apex of the flight; if the celestial spheres make music, then this was it. I heard the woodcock, but I did not see him, at least not then, in his lunar mating flight. I am sometimes envious of the night vision of certain nocturnal animals. The light-sensitivity of the fox's eye is increased by a reflective layer behind the retina called the *tapetum* (it is a reflection from the tapetum that makes a fox's eyes blaze red in the headlights of a car). Owls have huge funnel-like eyes so sensitive to night's faint light that they require a third eyelid, a kind of translucent shade, to diminish the dazzle of day. If I had the night vision of the fox or the owl, then I might have seen the woodcock in his lunar flight. But there is a way in which the disembodied music of an unseen woodcock is appropriate to the night, a kind of bodiless beckoning from an old wood. Even now I listen to the disembodied music of crickets and cicadas, the owl and the wind, night music, music from far off there among the stars. I once saw a wasp dragging the paralyzed body of a cicada up a tree. She was trying to get enough altitude to fly with her heavy booty back to her burrow. Sometimes I think that if I could only get enough altitude I could get back to where I started, to the old wood, to the old pool, to the cottage where thick-leaved branches sway and the stars rise like a mist on the eastern hedge, back even to the singular instant of Creation when bird and fish were flung with the galaxies into the wilderness of night.

❦

On summer evenings Scorpio burns in the southeast, and at the heart of the constellation is Antares, the red star, blazing like the eye of the fox. William Henry Smyth called Antares "fiery red." It is orange to my eye. The star is of the class we call red giants. Red giants are dying stars, bloated stars, rarefied, distended. Someday the sun will exhaust its energy resources and inflate to become a red giant. The Earth will be incinerated. The peach and the plum will shrivel on the branch. The five willows will be charred. The prized chrysanthemums will be burned to ash. The sun will swell and the wilderness that surrounds our hermitage will engulf us. Antares is a supergiant star. If Antares were where our sun is—93 million miles away—the Earth would be inside it.

In ancient China, our constellation Scorpio was called the *Azure Dragon* or *the Dragon of the East*. Burnham says of the Chinese Dragon: "[He] is not the hideous maiden-devouring monster of medieval Christian myth; he is the wise and majestic incarnation of the awesome power and infinite splendor of Nature." Jade carvings of the Azure Dragon are common artifacts in Chinese tombs, in the form of sinuous rods or scepters. These were called *ju-i*— the name literally means "as you wish." The curved tail of the constellation is hooked into the Milky Way. The Chinese imagined the Dragon to be emerging from the Celestial River, preparing to descend to Earth to confer great blessings on the sages who do him reverence.

Tonight, 1600 years after T'ao Ch'ien watched the Azure Dragon rise, I watched the rising of Scorpio. Huge, red-hearted Scorpio. *Ju-i*, the Dragon of the East, "as I wish." As *I* wish? Is this the answer to my prayers? The stars rise huge and indifferent. Their constancy insists upon their silence. Their distance ensures their deafness to my petitions. This is the indifferent wilderness, unbridled, encroaching.

There is a tendency for us to flee from the wild silence and the wild dark, to pack up our gods and hunker down behind city walls, to turn the gods into idols, to kowtow before them and approach their precincts only in the official robes of office. And when we are in the temples, then who will hear the voice crying in the wilderness? Who will hear the reed shaken by the wind? Who will watch the Galaxy rise above the eastern hedge and see a river infinitely deep and crystal clear, a river flowing from the spring that is Creation to the ocean that is Time? We are dust flicked from the scorpion's tail. The woodcock cries in ascending circles; the wild geese thrash the air with their heavy wings. The night is the old wood; the night is the native pool. Antares is a lamp, burning and shining; *rejoice in its light.*

"A dog barks somewhere in the deep lanes," wrote T'ao Ch'ien. "A cock crows at the top of the mulberry tree . . . Long lived I checked by the bars of a cage: Now I have turned again to Nature and to Freedom." I read T'ao's lines and I remember Roethke's prayer: *Ye littles, lie more close. Make me, O Lord, at last, a simple thing.* The stars tonight seemed close enough to touch. On summer evenings the Milky Way bounds the dooryard like a leafy hedge. Nebulas bloom like flowers. Vega, Arcturus, and Antares are hung like Chinese lanterns in the branches of the trees. *Night I embrace, a dear proximity.*

INDEX